The Current Economic Crisis

and

The Great Depression

The Current Economic Crisis

and

The Great Depression

Ashland
Community & Technical College
Library

College Drive Campus

Philip S. Salisbury

Copyright © 2010 by Philip S. Salisbury.

Library of Congress Control Number:		2010910317
ISBN:	Hardcover	978-1-4535-3827-2
	Softcover	978-1-4535-3826-5
	Ebook	978-1-4535-3828-9

All rights reserved. No part of this book may be reproduced or transmitted in any form or by any means, electronic or mechanical, including photocopying, recording, or by any information storage and retrieval system, without permission in writing from the copyright owner.

This book was printed in the United States of America.

To order additional copies of this book, contact:
Xlibris Corporation
1-888-795-4274
www.Xlibris.com
Orders@Xlibris.com
83474

TABLE OF CONTENTS

Title	Page
Preface	17
Introduction	19

Part I

Chapter I 23
    Implications For 2007
    and Beyond

Chapter II 32
    Looking Ahead

Chapter III 44
    An Historical Overview:
    Pre-1929 To 1940

Chapter IV 92
    A Conceptual Overview

Chapter V 95
    Schumpeter and
    Capitalist Evolution

Chapter VI 149
    WWI and Gold

Chapter VII 162
    Keynes and Effective
    Demand

Chapter VIII 170
    Money Supply and
    Banking

Chapter IX 185
    Farming

Part II

Chapter X 193
    Theses and Some Tests

Chapter XI 204
    Labor Force and
    Aggregate Demand

Chapter XII 216
    Explaining the
    Great Depression

Appendix I 237
    A Glossary of
    Abbreviations

Appendix II 243
    An Aggregate
    Demand Equation

References 247

Tables

Table 1.1 27
    Number of Males
    and Females with
    Income by Income
    Category and Mean
    Income: 2006

Table 3.1. 62
    Labor Market Statistics
    1925-1940

Table 3.2 72
   Farms and Types of
   Farm Ownership

Table 3.3 83
   Homicide Rates, Suicide Rates,
   And Unknown Causes:
   Death Rates 1929-1940

Table 4.1 94
   Business Cycle
   Reference Dates

Table 5.1 124
   M2, GNP in Current
   Dollars, and GNP
   Production
   Per Man-hour

Table 5.2 136
   Banks Suspended
   and Amount of Suspended
   Deposits

Table 5.3 137
   Bond Rates 1932-1935,
   Wholesale Price Index,
   and Consumer Price Index

Table 6.1 150
   Unemployment and the
   Unemployment Rate in
   the U.S.: 1916-1940

Table 6.2 152
   U.S. Gold Stock, Annual Change,
   and Personal Income

Table 8.1 170
   Correlation Matrix

Table 8.2 173
   Civilian Labor Force and
   M2/Person in CLF

Table 8.3 175
   All Banks,
   Non-national Banks
   And the Percent of
   Non-national Banks

Table 8.4 177
   Suspended U.S. Banks
   and Suspended Non-National Banks

Table 8.5 180
   Velocity of Money, M2,
   GNP, and Adjusted
   Unemployment Rate

Table 8.6 183
   Correlation
   with Loans

Table 8.7 183
   Correlation Matrix
   CPI, GNP, and M2

Table 8.8 184
   Consumer Price Index,
   1920-1940 (1914 = 100)

Table 9.1 186
   Non-Farm and Farm
   Employment

Table 9.2 189
   Farms and Types of
   Farm Ownership

Table 10.1 197
   Correlation Coefficient (N=120)

Table 10.2 198
   Unemployed Adjustment
   for Discouraged Workers
   1930-1940

Table 10.3 200
   Correlation Matrix (N = 41)

Table 11.1 204
   Labor Market Statistics
   1925-1940

Table 11.2 205
   Adjusted CLF, Adjusted
   Unemployed, Number Unemployed,
   and Number of Discouraged Workers

Table 11.3 206
   Adjusted CLF, Adjusted
   Number Unemployed,
   Adjusted Unemployment Rate

Table 11.4 208
   PC, GOV, GDPI,
   Z, and, AD

Table 11.5 213
   Correlation

Table 11.6 214
   Adjusted Unemployed, Personal Income, and
   Percent Change in Personal Income

Table 12.1 222
    Age and Percent Owning Their
    Own Home

Table 12.2 223
    Correlation Matrix

Table 12.3 226
    Correlation Matrix

Table 12.4 230
    Banking Reserves in the U.S.
    1930-1940

Table 12.5 232
    Personal Income
    in Current Dollars:
    1920-1940

Figures

Figure 1.1 24
    Annual Live
    Births: 1900-2004

Figure 1.2 26
    Civilian Labor Force 45 Years
    Old Compared to CLF 64 Years

Figure 1.3 28
    Average Incomes
    of Males and Females
    in Different Age Categories

Figure 2.1 33
    Sales of Stock
    Compared to the
    Adjusted
    Unemployment Rate

Figure 3.1 70
    Farm Acreage per Farm
    Employee

Figure 3.2 71
    Farm and
    Non-agricultural
    Employment

Figure 4.1: 93
    Stock Sales Compared
    to the Inverse of the
    Unemployment Rate

Figure 5.1 123
    M2, Productivity Per
    Person-hour, and Gross
    National Product

Figure 5.2 125
    Gross Nation Product, divided
    by M2 = (Velocity)

Figure 5.3 136
    Number of Banks
    Suspended and Deposits
    Suspended

Figure 5.4 142
    U.S. Non-agricultural Employment
    and Farm Employment

Figure 6.1 150
    The Number Unemployed
    and the Unemployment Rate
    in the U.S.

Figure 6.2 155
    Gold Stock in the U.S.
    and Annual Change

Figure 6.3 158
    Gold Stock Compared to the
    Consumers' Price Index

Figure 6.4 159
    Gold Stock vs. Personal
    Income in Current Dollars

Figure 8.1 172
    M2, Gold Stock, and
    U.S. Currency
    in Circulation

Figure 8.2 176
    All Banks, Non-national
    Banks, and Percent
    Non-national

Figure 8.3 178
    M2 and the Adjusted
    Unemployment Rate

Figure 8.4 179
    M2, GNP in Current
    Dollars, and Velocity

Figure 8.5 181
    Deposits, Reserves and
    Deposit-Reserve Ratio

Figure 9.1 185
    Non-farm and
    Farm Employment

Figure 9.2 187
    Farm Acreage per
    Employee and Income
    Per Farm

Figure 9.3 188
    Index of Farm Output
    1929-1940

Figure 10.1 199
    Personal Income and the
    Inverse of the Adjusted
    Unemployment Rate

Figure 10.2 202
    M2, Personal Income,
    and Gold Stock

Figure 11.1 212
    M2, Personal Income,
    and the Adjusted
    Unemployment Rate

Figure 12.1 216
    The Annual Number
    of Immigrants, 1900-1940

Figure 12.2 218
    Live Births Compared
    to All Immigrants:
    1900-1940

Figure 12.3 219
    All Immigrants
    Lagged 16 Years
    Compared to Live Births
    Lagged 16 Years

Figure 12.4 220
    The Inverse of the
    Adjusted Unemployment
    Rate Compared to
    Immigration Lagged 16 Years

Figure 12.5 221
    All Immigrants Lagged
    16 Years vs.
    Housing Starts

Figure 12.6 223
    Immigrants Lagged
    16 Years Private Housing
    Starts, and Non-farm Foreclosures

Figure 12.7 224
    Immigrants Lagged
    16 Years Compared
    with Private Residential
    Costs

Figure 12.8 235
    Personal Income
    and Bank Reserves
    1900-1940

To My wife and partner—Mary

Preface

This book had its origins in data-induced curiosity about the decline in the number of foreign born males of working age in 1930 and 1940. Finding these numbers involved a circuitous search that uncovered sex and age related historical data from 1900 to 1940. Graphing that data for male foreign born men of working age showed immigration peaks in 1907 and 1913-1914. An assessment of the early male immigration data indicated an average age of approximately 40-45 years. This led to exploration of the hypothesis that these immigrant males of working age in the 1907, 1913-1914 periods had lived out their natural lives, as well as their working lives.

The variety of views available provided sufficient reason for continued research on the depression phenomena. What followed, was the identification, data-entry, and sometimes transformation of data sources into two major data files. Graphic analyses, descriptive, and multivariate analyses were conducted. An application of Stress Theory was developed with a disaggregation of personal income, disposable personal income, or personal consumption making it possible to understand how population could affect aggregate demand.

The foremost question facing economists and policymakers currently is whether their will be a repetition of similar events in the 2007-2020 period? Martin Feldstein, in his edited volume *The Risk of Economic Crisis*, states that though the United States has avoided major economic collapse in recent decades (approximately 1950-1990) "there is no doubt that the possibility of another such financial and economic crisis cannot be dismissed" (Feldstein 1991). The reader may find it interesting to determine whether the current volume offers any new information to help resolve that issue.

The ideas that evolved were strengthened through weekly luncheon discussions I had with my colleague, Professor Emeritus Phillip M.

Gregg, Ph.D. The repartee served to sharpen the ideas contained herein. Most importantly has been the love and support of my wife and partner Mary.

....

A data base is available on CD. It contains data (371 variables & 248 cases) from the 1900 to 1940 period used in the analytics present in this book.
The CD is formatted for Statistica.sta databases. The CD can be obtained by mailing 18 dollars (checks only) with your return address to:
Economic & Population Trends
Depression CD
3004 Arlington Drive
Springfield, IL 62704

Introduction

A question central to economic thought is: how can fluctuations in economic activity be explained? Outside of theories depending on "shocks" to the economies of developed economic systems, economic theory has not been able to provide explanations of business cycles and economic downturns, which allow predictive or explanatory insight into their genesis and maintenance.

Economists are not willing to dismiss the possibility of another economic crisis such as that experienced in the 1930's (Feldstein et al., 1991, p. ix). The economic boom experienced after 1992 certainly allayed such fears on a temporary, and perhaps on a more permanent basis. What is disturbing, however, is our continued lack of understanding of the forces that create economic downturns as well as growth. Four major currents of economic thought arrive at differing interpretations of similar phenomena, but admit to the continued likelihood of disequilibria in the economies of advanced nations.

The study of business cycles excludes changes in business conditions occurring in periods known as "crises" (Mitchell 1927). The stock market crash and the period of extended economic downturn (1929-1940) do not qualify as a business cycle. For that reason, the study of the stock market crash in 1929 and the following period of economic contraction and economic downturn have survived as a topic of specialized scholarship.

There have been efforts to provide partial theoretical explanations as well as data-based explanations for the events of 1929-1940. These efforts have resulted in intensive efforts to pour over events of the stock market crash and the subsequent downturn in detail for possible explanations.

Many economic explanations of events rely on factors of change that will eventually bring the economy to a state where all forces, including business and employment are relatively fully utilized. The search for equilibrating forces that occurred in the 1930's has not, as of yet, turned up these forces. This book will suggest that there has been a failure to empirically examine some of the most important of these forces for change.

This manuscript will be divided into an initial chapter setting forth the author's views on the underlying dynamics that are prompting the current economic crisis. The second chapter acts as a summary of the remainder of the book and acts as a prelude to the remainder of the book. The third chapter presents an overview of the historical events of the Great Depression era. Chapters presenting major views that are used to explain the stock market crash, the Great Depression (1929-1940), its advent, and its duration follow. These views are, a historical description of the events leading to the stock market crash in the United States, a historical conceptualization of manias, panics, and stock market crashes, Schumpeter's view of economic innovation and creative destruction, a World War I and gold standard explanation, a demand theory explanation, and an explanation based in monetary theory. Evidence consisted of the monetary theory explanation, a presentation of the impact of immigration, evidence supporting the demand theory explanation, an analysis of inflation and deflation, and a view of what was happening to farming. The author makes use of Stress Theory and a set of theses to integrate three of the theoretical approaches. These understandings will assist in clarifying the discussion and documentation that preceded it.

As will be demonstrated in the following pages, there are many diverse perceptions about the causes of the Great Depression. It is difficult to sort out what factors, events, and variables are causes of the Great Depression and which ones are descriptive of it. Some individuals have asserted that their explanations are causal. while they in fact describe the conditions and events prevalent during the depression era in the United States.

As a result, some of the past perceptions of why the Great Depression era occurred are clarified. Additionally, the pervasive nature of the Great Depression in the United States becomes evident as the text progresses.

Part I

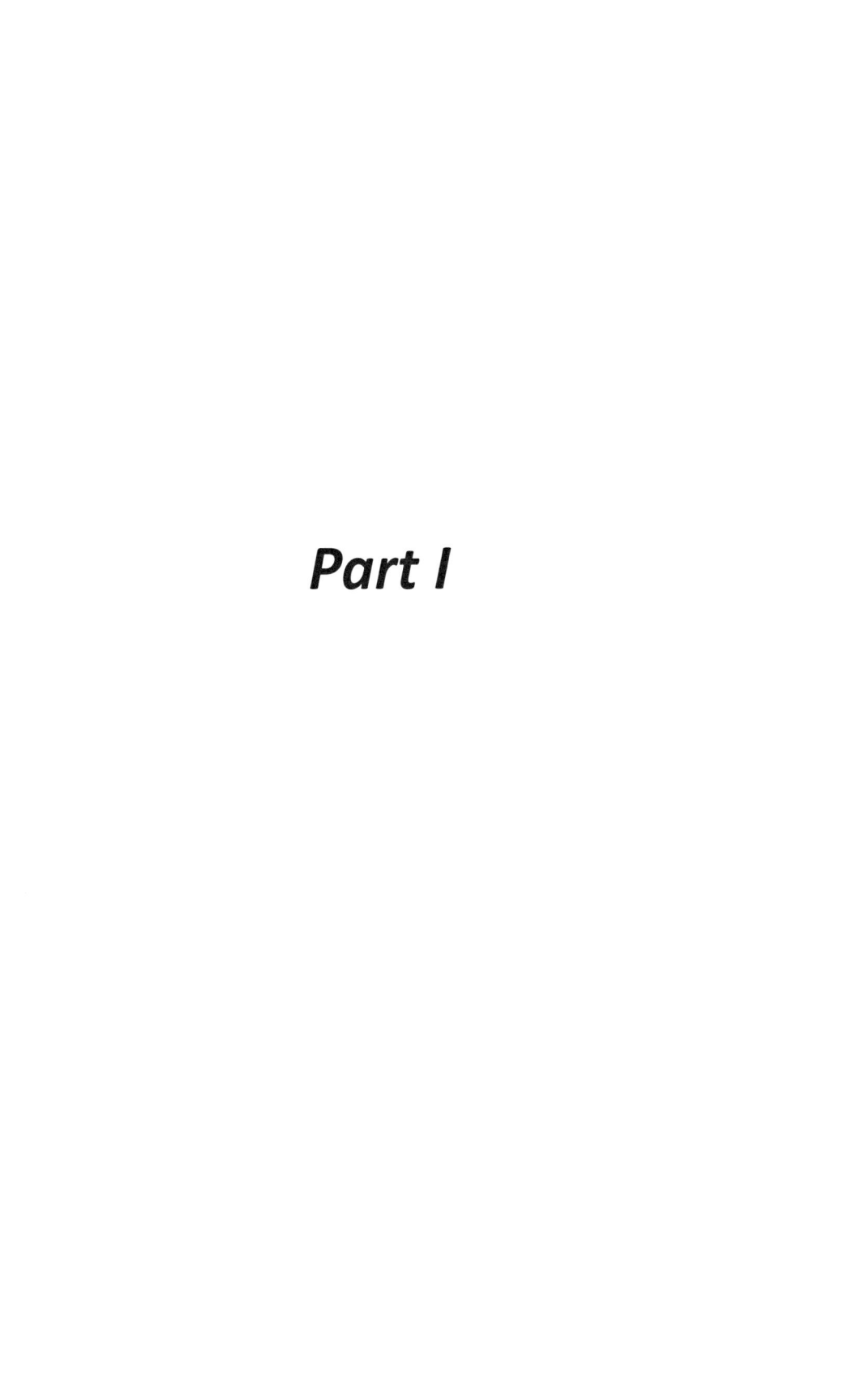

CHAPTER I

IMPLICATIONS FOR 2007 AND BEYOND

The economic circumstances in the United States now and during the 1929-1940 period have several elements in common. There were no bank failures in 2005 and 2006. There were three bank failures in 2007, there were 10 bank failures (to August 2008). The failure of Lehman Brothers on September 14, 2008, the decline of the Dow on September 16-17, 2008 by over 949 points, a housing crisis, a credit crisis, increasing unemployment, and declining confidence in the U.S. economy are factors that, although they are currently not as severe as the Great Depression, raise concerns about future potential decline.

Underlying the Great Depression Era were nine main dynamics. These dominant dynamics were variations in M_2 level, variation in bank reserves, a housing boom and bust, foreclosures, unemployment, lack of faith in financial institutions, liquidity followed by illiquidity, deflation, and declining personal income. Secondly, the increase (1929-1933), decline (1937), and then increase (1938-1939) in the adjusted unemployment levels showed the evolving human impact of the Great Depression. Included in the number of unemployed is the population dynamic of the declining numbers of working age male immigrants. Over the period from 1930 to 1940 the immigrant decline was not being offset by equivalent numbers of immigrant, working age males. From 1931 through 1940, the annual number of immigrants was less than 100,000.

The period from 2007 to 2029 is affected by a different, though somewhat similar, dynamic. The dynamic is the aging of the 45-64 age and sex groups and these group's subsequent declining incomes. The decline and leveling of the 45 year age group in 2016 to 2021 is an important low point which could mark another period of low effective demand.

The purpose of this chapter is to demonstrate how the aging live birth data by age and sex group and application of income data can be used to roughly characterize the current and future effects of population subgroups on the U.S. economy.

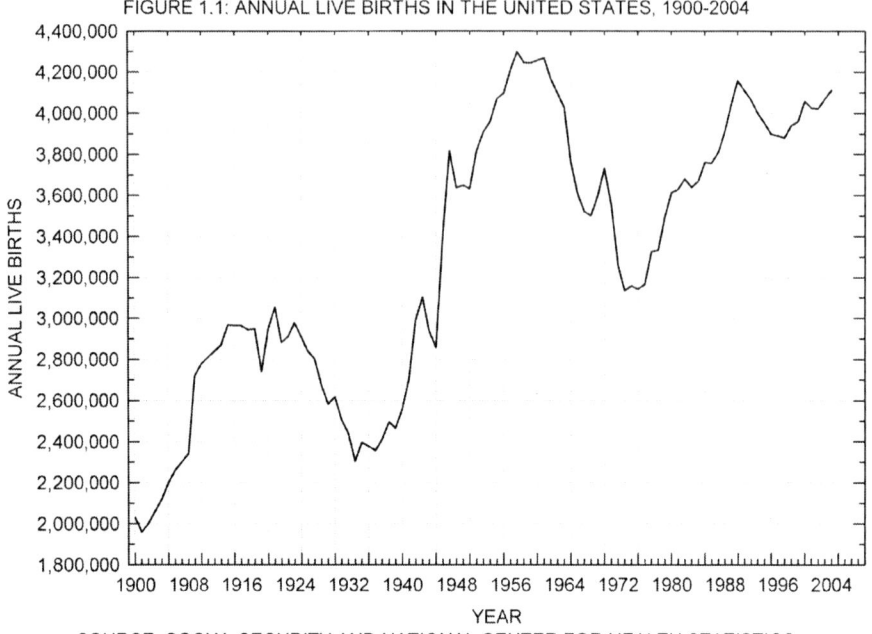

FIGURE 1.1: ANNUAL LIVE BIRTHS IN THE UNITED STATES, 1900-2004
SOURCE: SOCIAL SECURITY AND NATIONAL CENTER FOR HEALTH STATISTICS

There was a continual growth in the number of live births in the U.S. between the low in 1933 and the peak from 1958-1961. This was an increase from 2.304 million live births in 1933 to 4.268 million in 1961. In the 1933 to 1961 period 97.488 million live births occurred. From there the annual number of live births declines from 4.268 million live births in 1961 to 3.144 million in 1975. That was a total of 54.54 million live births from 1961 to 1975.

In comparing these numbers, it becomes clear that the decrease in live births from 1961 to 1975 represents a major force influencing the contraction of the economy. The decline of the population is currently in process and will be for some time. This represents a significant demographic effect on personal income and personal consumption in the economy.

Why are these increases and decreases important? First, each year the number of surviving live births in a cohort ages. The number in the cohort declines from year to year. Secondly, when each year of live births reaches employable age, the earnings of the age group vary as the cohort ages. I will first examine the effect of live births aging. The number of live births from a given year can be multiplied by the survival rate of the cohort at a given age. This results in the number of individuals from a cohort remaining at given ages.

(Live births/year)(cohort survival values y_0 to y_n) = number surviving, t_0 to t_n.

Second, only a portion of the persons surviving to a given age will participate in the labor force and be employed, or have a source of income. The labor force participation rate is used to determine the number in a population in the civilian labor force (CLF). The number in the CLF can be divided into three groups; those employed, those unemployed and searching for work, those unemployed and not searching for work.

An illustration of this situation is provided in Figure 1.2. There are three principles to remember. First, if a year of live births is lagged a given number of years each datum of live births is the age of the lag. If a cohort is prepared from a year of live births it may be multiplied times each data point on the survival curve. The resulting numbers represent the number surviving at that time. Figure 1.2 characterizes surviving live births lagged 45 years in the civilian labor force and surviving live births lagged 64 years in the civilian labor force versus the year in which the lagged data point occurs.

What is of note about Figure 1.2? Between the 45-year-old civilian labor force line and the 64-year-old civilian labor force line there are lags of 46, 47, 48, 49, 50, 51, 52, 53, 54, 55, 56, 57, 58, 59, 60, 61, 62, and 63 years which are not shown. Thus, the lagged 45 and 64 year old lines are only being used to represent all those lines between 45 and 64. In the year 2006 and beyond to 2016-2021 the number of individuals declines to a four year trough. Where are the 45-64-year olds going? They are aging as well as dying. The number of individuals becoming 64 and over grows old and eventually declines (not shown in Figure 1.2).

Every lag of live births from 45 to 64 needs to be multiplied by the survival rate for a given age and a given age level's appropriate sex. The surviving 45 & 64-year olds have been multiplied by their survival rates. Second, each line of lagged live births is multiplied by its labor force participation rate to result in the number in the civilian labor force. What conclusions can we draw from this graph (Figure 1.2) and supporting data in Table 1.1?

FIGURE 1.2: CIVILIAN LABOR FORCE 45 YEARS OLD COMPARED TO CIVILIAN LABOR FORCE 64 YEARS OLD

SOURCE: FROM LIVE BIRTH DATA (NCHS), SURVIVAL RATES (SOCIAL SECURITY) AND LABOR FORCE PARTICIPATION RATES (BUREAU OF LABOR STATISTICS)

Table 1.1: Number of Males and Females With Income by Income Category and Mean Income: 2006*

Age Category	Number of Males (000)	Mean Income per person (males)	Total Male Income (billions)	Number of Females (000)	Mean Income per person	Total Female Income (billions)
15-24	14,093	$15,213	$214.397	13,267	$12,290	$163.051
25-34	19,045	32,131	611.935	17,151	29,693	509.265
35-44	20,374	42,637	868.686	19,199	34,762	667.396
45-54	20,247	45,693	925.147	20,135	36,740	739.730
55-64	14,905	57,219	852.849	15,073	33,289	501.765
65-74	8,417	39,435	335.868	9,781	21,588	211.152
75 & over	6,547	30,623	199.750	9,975	17,990	179.450

*Current Population Survey, Annual Social and Economic Supplement (ASEC), U.S. Census Bureau, *http://pubdb3.census.gov/macro/032007/ perinc/new01_019.htm*, Accessed March 23, 2008, 2006 Income data is from 2007 ASEC

Figure 1.2 shows the decline in the civilian labor force (CLF) for a 45-year lag compared to the rise in the 64-year-old CLF (2007-2034). The correlation between the two CLF lines is r= -.76 (N=28, p=.000). The total in the 45-year-old CLF line from 2007 to 2034 is 84.89 million persons. The overall number in the 45-year-old CLF from 2007 to 2021 is 44.79 million individuals.

Continuing the analysis of Figure 1.2, from 2007 to 2018, the 45-year-old CLF declines a sum of 36.82 million persons. From 2018 to 2021, the 45-year-old CLF approximately levels. The sixty-four year old CLF increased by a sum of 38.51 million persons from 2007 to 2025. From 2007 through, 2033, 54.50 million 64 year old CLF persons became 65 years or older. The 64 year-old line represents approximately 16.35 percent of the current U.S. population. From 2011 to 2029, 64-year-olds outnumber 45-year-olds by 14.61 million individuals.

Figure 1.2 shows the decline in the CLF for 45-year-olds. There is a roughly reciprocal incline in the number of live births in the in the 64-year lag from 2015-2021. Thus, the departure of these people from income earning status will have a major impact on the economy. This can be explained by two major factors. First, the decline in the absolute numbers in the 45-year-old lagged live births (really 45-54-year-old males and females) means that there will be fewer income earners at those ages. The peak and then declining incomes of males 55-64 and above have a significant impact on the U.S. economy as they pass into the 65 and over age categories.

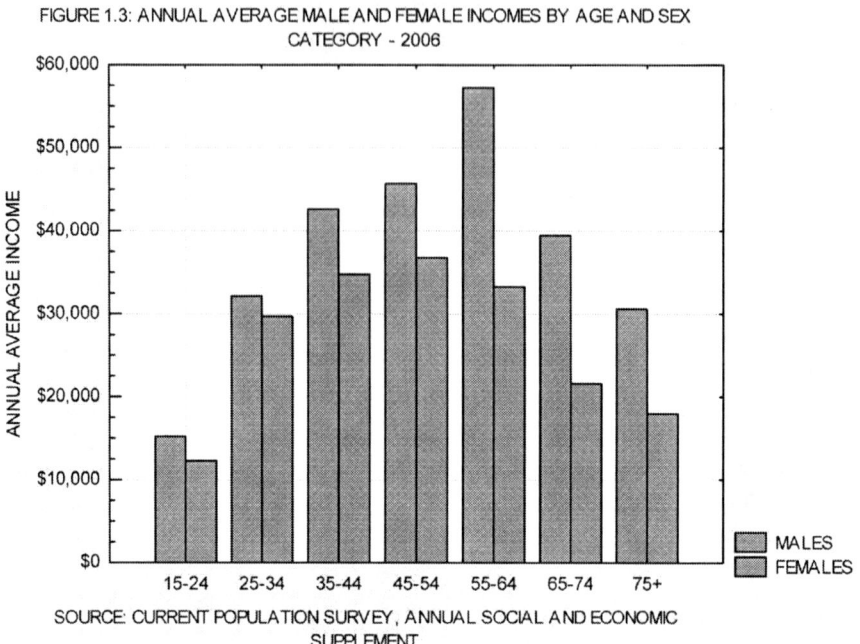

Figure 1.3 and Table 1.1 give us evidence from which to draw conclusions about the current dynamics of population and income. First, Figure 1.3 and Table 1.1 demonstrate how mean income differs for different age and sex groups. Based on what Table 1.1 and Figure 1.3 reveal, the following observations can be made:

- Male income increases from a mean of 15,213 dollars for the 15-24 age category to a maximum of a mean of 57,219 dollars for the 55-64 age group. Average incomes decline from that peak to an average of 30,623 dollars for those 75 and above.

- Female mean income increases from a low of 12,290 dollars for the 15-24 year age category, to a peak of 36,740 dollars for 45-54-year-olds. Mean income declines from that peak to 17,990 dollars for those 75 years and older.

- In Figure 1.2 the lagged CLF line for 45 year old starts declining from 2007. The decline plateaus in 2018-2021 and begins a gradual increase from 2022 to 2024.

- When interpreted, Figure 1.2 means that both the 45-54 female year old group and the 55-64 year old group are aging. These are respectively the two highest average income groups for both females and males. These age and sex categories are aging and total incomes will decline accordingly. To determine the effects of aging on both the 45-54 year olds and those aged 65 and over, civilian labor force, employment estimates, and average incomes would be needed.

- In summary, the five weeks (September 7 through October 14, 2008) have witnessed collapsing financial institutions (Bear Sterns, Fannie Mae, Freddie Mac, and AIG, Washington Mutual Inc), and the adoption of a rescue package. In one week the stock market (Dow) dropped about 1,580 points. Several important conclusions can be drawn:

- There will continue to be more volatility in the stock market (NYSE) and the DOW.

- The decline in the number of individuals in the four groups, with the highest average incomes as they age, will persist from 2008-2021 and possibly beyond. This 14-year period will bring depression-like conditions to the United States. Even with appropriate policies, the depression may only be moderated. Persistent unemployment will continue through the 2008-2021 period as seniors age into the 65 and over age group.

- It is clear that this is both a housing crisis and a credit crisis. The source of the housing crisis was that housing was overbuilt. In other terms, the number of individuals at house purchasing age has been in decline and will continue to decline. Of all homebuyers, 34 percent are ages 45-64 (National Association of Realtors, 2007). The credit crisis has occurred because of the bad mortgages and loans written. Bundling of mortgages, creating securities out of them, and insuring the securities when the sources of the securities cannot be traced created a hazard in the stock market.

- Banks became reluctant to loan money to others because of a lack of confidence in the borrower's ability to repay.

- Banks do not have the assets to loan because the number of depositors will decline as the 45-64 year age and above age group declines. As the number of people 45 and over declines, assets will be used in retirement, and pension funds will decline in value.

- In these circumstances, the numbers of people in certain age groups will decline. Under these conditions of declining age group numbers, and declining incomes, total personal income and total personal consumption will decline. Effective demand will decline and the economy will experience deflation (This tendency towards deflation may be offset by the large amount of M_2 produced and what have been significant government expenditures).

- The surviving live births lagged 45 years from 2004 to 2029 will age out of the 45-year-old lag category and the 64-year-old lag category will age into the 65-year-old and above age group. This estimated level of change in the CLF for those respective ages will affect the actual number of individuals, male and female surviving to the dates on the X axis of Figure 1.2. By applying survival rate, labor force participation rates, and unemployment rates more accurate numbers are obtained.

In summary, this analysis deals only with the CLF effects from 2007 to 2021. Later effects for the years beyond 2021 can be studied to gain additional insights. Government assistance as a last resort has come into play. The mating habits of U.S. citizens are subject to the influences of history. The likelihood that we cannot control levels of live births leaves us at the mercy of this powerful demographic force. The U.S. economy currently faces a major problem due to the lack of effective demand (Keynes 1936). Resources used by the government need to be directed at that problem.

CHAPTER II

LOOKING AHEAD

The Historical Overview (Chapter III) and the Conceptual Overview (Chapter IV) provide a detailed account of circumstances prior to 1929, and the years that followed. Kindelberger (1989) provides an excellent framework within which to generalize the course of financial crises. His work provides a good descriptive framework for looking at the period prior to 1929, 1929 itself, and the collapse of the New York Stock Exchange (NYSE). There are some who would argue that the events of the stock market Crash in October 1929 and what followed were a direct result of the Crash. The evidence for this view is in Chapter IV (Figure 4.1).

Figure 2.1 shows that the Crash and the period from 1929 through 1932 featured a steep decline in the dollar value of shares traded. Whether this contributed to a lack of confidence in U.S. financial institutions that manifest in banking panics, bank failures, and bank suspensions cannot be empirically determined. Figure 2.1 shows the adjusted unemployment rate lagging the decline in stock sales. Secondly, this book makes it clear there were other factors in operation as well.

FIGURE 2.1: SALES OF STOCKS COMPARED TO THE ADJUSTED UNEMPLOYMENT RATE

SOURCE: CALCULATIONS FROM HISTORICAL STATISTICS OF THE UNITED STATES DATA

Schumpeter

Schumpeter (1938) contributes a unique perspective to the interpretation of business cycles. His concepts of economic evolution and innovation provide a useful way of thinking about technological change. While he argues against calling such change progress, he makes an excellent case for the existence of a process of evolving growth, maintenance, and decline of innovations. He sees this process as the core of capitalism and capitalist evolution.

Schumpeter's discussion of the terms *equilibrium* and *near equilibrium* establish his view that an economy rarely reaches equilibrium. It was through these discussions that the term *economic evolution* came into being.

Schumpeter makes an effort to use a tripartite framework of cycles or waves. Here waveforms titled Kondratieff (fifty-four years); Jugular (nine years); and, Kitchin (four years) are used to assist in his historical interpretation of economic evolution and business and industrial events.

When it comes to the interpretation of the period from 1929 through 1938, however, Schumpeter deviates from his periodic interpretation of economic evolution. Seeing the economic process during that period, he had a difficult time adjusting the concepts of economic evolution to fit the period's events. He became skeptical of what was happening and questioned whether the U.S. economy was evolving into something other than capitalism.

At the end of the second volume of his work, he suggests that the Roosevelt administration was populated by administrative neophytes who reacted to their situation out of a lack of experience. He contrasted the administrators under Roosevelt unfavorably with the administrators of the British civil service.

Second, he saw the backers of the Roosevelt administration as unfavorable to business. He saw this anti-business attitude as a major change in public opinion, which would result in a changed style of government and a sociological drift that he thought could not be expected to change. Schumpeter thought that if his schema were followed, prosperity and recovery would be more strongly delineated and recession and depression less so in the next 30 years.

The World War I and Gold Standard Hypothesis

World War I, Temin (1989) maintains, strained the gold standard, leading to its suspension, though not its termination. The inter-war gold standard was unstable through World War I. After World War I there was a move in the United States Great Britain, France, and Germany to revive the gold standard in effect prior to the war. Temin sets forth the idea that the Depression was not a necessary consequence of World War I. After the war's disruption, efforts to reestablish the gold standard were important in generating the Depression. Its beginning was not unavoidable in 1929. If policymakers had not failed to free themselves from their adherence to the gold standard, they could have enacted corrective policies.

Temin suggests that the presence of the gold standard dictated inflation and deflation. Deflationary forces were becoming stronger in

1929. Maintaining the gold standard in an industrial economy further exacerbated deflationary tendencies (see Chapter V, Table 5.2).

The British were several steps ahead of the United States. Britain abandoned the gold standard on September 20, 1931. The shift in policy allowed the British economy to lower export prices relative to foreign prices. It also allowed the British to establish a more relaxed monetary policy. The Bank of England did not allow gold to flow out of its country and thereby built up its reserves. These actions had a deflationary impact on the rest of the world.

It is Temin's view that if going off the gold standard had been done earlier than January 1934 in the United States (U.S.) the deep deflationary period from the latter part of 1929 through the depths of deflation (1933) and after could have been prevented. After the U.S. rejected the gold standard, the U.S. gold stocks began to increase at a rapid rate.

Eichengreen and Sachs (1985, 1986) view the interwar gold standard as the principal carrier of deflation. They conclude that devaluation or abandonment of the gold standard was a first step in national and world recovery. In their research (1985), they examined differences between countries abandoning the gold standard at an early time (at or before 1931) and those countries abandoning the gold standard later. They concluded that countries leaving the gold standard early had much more rapid recoveries than those remaining on the gold standard. Consistent with their perspective was that the notion that the gold standard led to a monetary contraction.

Eichengreen and Sachs (1985) tested a hypothesis about the role of wages in determining aggregate supply. using a sample of ten industrial countries in 1935 with the output variable industrial production they find a strong negative relationship between output and real wages across the ten countries. Countries that fell into the high-real-wage and low-output category were countries that remained on gold well after 1931.

Bernanke (2000) uses a larger sample (N=22, 1931-1936) than do Eichengreen and Sachs (1985, 1986) to test the view that leaving the gold standard gave nations a greater ability to increase the money supply. The view presented is that staying on the gold standard promoted the deflationary process. Bernanke (1995, 2000) advances the view that an essential first step out of the Depression to national and global economic recovery was to go off the gold standard.

Countries included in the high-real-wage, low-output area remained on gold beyond 1931. These nations included Belgium, France, Italy, and the Netherlands. Nations with low real wages and higher output included Denmark, Finland, Norway, Sweden, and the United Kingdom. Bernanke is not able to explain this phenomenon.

Bernanke also finds evidence of nominal wage stickiness in the Depression and indicates that sticky wages were a dominant source of non-neutrality.

Based on more extensive data while reinforcing the view of Eichengreen and Sachs (1985, 1986) leaves Bernanke to state:

> . . . we are not aware of any plausible story of why these declines in spending should have affected so many disparate countries around the world so nearly simultaneously and in particular why they should have been more persistent in countries remaining on the gold standard. (p. 286)

The Money Supply Hypothesis

Friedman and Schwartz (1963) see the "Great Contraction" as a liquidity crisis prior to and during which there was a lack of a response by the Federal Reserve, specifically to purchase government securities. The Governor of the New York Bank, Benjamin Strong, is viewed by Friedman and Schwartz (1963, pp. 412-13) as a leader who was willing to use the power of the Federal Reserve to halt any panic should it arise. They view the death of Governor Strong in October 1928 as an essential factor in the subsequent failure of the New York Bank to address the liquidity issues that arose. Strong had the support and backing of other financial leaders in addition to the personal courage and force to assert his views. With Strong's death, there was no forceful voice to assert the need for the Federal Reserve Banks to meet the liquidity crisis.

Irving Fisher (1934, p. 151) believed that if Strong had lived, "we might have had the stock market Crash in milder form, but after the Crash there would not have been the great industrial depression."

I do not agree with this position, as it is a conjecture about the actions of Governor Strong and assumes a single factor at the base of the "Great Contraction." Not only was there a monetary imbalance but there was an aggregate demand and supply imbalance due to the decline in personal income levels resulting from high levels of adjusted unemployment, deflation, and perhaps declining M_2 levels (supported by Equations 11.5 and 11.6 in Chapter XI). Furthermore, declining immigration following the 67th Congress's action limiting the immigration of aliens into the United States (May 19, 1921) and extension of that Act on May 11, 1922. The low levels of immigration after the stock market Crash until 1940 (under 100,000 annually from 1931-1940), were indicative of decline. As Chapter XII suggests both parts of M_2, a paucity of central bank money and the decline in deposits (a source of checkbook money) contributed to the liquidity crisis.

Monetary theory would benefit from the application of Stress Theory to the monetary Equations developed by Fisher (1906) and Friedman (1956). To do so, would be to regard each Equation as a stress Equation or an Equation of equilibrium or disequilibrium.

Effective Demand

If further data were available and included (average personal income for each sex, age, foreign-born ethnicity, and education subgroup status) it would be possible to estimate the loss in personal income resulting from the variation in lagged immigrants. Since this is not possible, the next best alternative, using available data, is to lag existing immigrant data assuming an average age of 33-34 years upon arrival into the U.S. Immigrant data from 1900 to 1914 covers a peak period of immigration into the U.S. The immigrants are lagged 16 years and the survival rate for males aged 49 years are applied to the original number of immigrants lagged. A curve for the variation of immigrants age 49 years is obtained. Keynes' (1937) concerns about the economic impact of declining populations is applied to this result.

Keynes (1937) warned that a significant reduction in population could affect consumption levels in a society. The loss of immigrants over the 1930 to 1940 period is a significant reduction in population. A reduction in personal income attended this decline. As has been clearly explained in this book, a major reduction in either personal income or, consequently, personal consumption will result in declines in aggregate demand. Business and industry will respond to this contraction in demand by reducing supplies, thereby increasing unemployment.

In the view of the author, this was an important mechanism at work in the Great Depression era. It is not possible, given the distribution of the data, to make any inferences about whether this process began as early as 1929, thus explaining the Crash.

As a contrast, this book suggests that total personal income in the U.S. shrank partially in response to the declining magnitude of selected white foreign immigrant population age groups in the 1929-1940 period. More importantly, a housing boom was created in 1922-1929. This occurred during the peak years of the immigrants lagged 16 years. The housing boom in 1922-1925 was followed by a housing bust in 1929-1933. Housing starts in 1929 were followed by an increasing number of foreclosures. The foreclosures peaked in 1933 and then declined.

Of greater importance are the relationships of total unemployment, and the wholesale price index to personal income (see Chapter XII, Path Diagram 2). The period from 1930, to and including 1933, marked a period during which bank panics occurred. Wicker (1996) contrasts the 1930-1932 period to 1933. The 1930-1932 period consisted of a number of regional crises. The 1933 banking crisis was a nationwide period of general lack of confidence in the banking sector as indicated by a decline in total bank reserves. The decline in reserves was a result of increasing unemployment, a declining wholesale price index, and declining personal income. The decline in personal income resulted in the diminishing of checkbook money supplied by depositors. The decline in personal income and checkbook money combined with the Federal Reserve's failure to supply sufficient amounts of central bank money to banks resulted in bank failures and reinforced the pattern of credit rigidity after 1929.

Wicker

Wicker (1996) in his *Banking Panics of the Great Depression* divides banking crises into those that occurred in 1930 through 1932 terming them regional crises and the crisis in 1933 which occurred throughout the United States. Wicker (1996) sees money stock as an important part of the banking crises from 1930-1933 (U.S.). He (p. 58) maintains that increased demands for currency relative to demand deposits can be explained by the lagged effects of the continued erosion of confidence due to the collapse of Caldwell and Company. Increased hoarding resulted from these disparate banking crises in three Federal Reserve Districts). In 1933 depositors became vulnerable to additional shocks to the banking system.

The Michigan banking holiday on February 14, 1933, initiated a rash of bank moratoria, that lasted approximately three weeks. Bank holidays in one state were quickly transmitted to adjoining states. The anticipation that restrictions on withdrawals would be enforced in their states motivated depositors to reduce balances in their accounts. A contagion of closures developed. The spread of panic was accompanied by bank moratoria throughout the U.S. While the bank moratoria were not a solution, they bought time for the banking officials. State banking officials were given authority over the banks. No confidence remained in the Federal Reserve, the RFC, and the President. Banks in all 48 states either had restricted deposits or were closed when Roosevelt was inaugurated. He declared a national bank holiday on March 6, 1933. The proclamation had the effect of codifying what was already occurring. There were two underlying considerations. The first was the need for national leadership. Secondly, the external drain of gold threatened the convertibility of the dollar into gold (particularly the New York Bank). The proclamation of the holiday transferred the power to open closed banks from the states to the federal government. The national bank moratorium began on March 6, 1933 and had an indefinite end date.

The Emergency Banking Act of 1933 gave the U.S. government the powers to reopen banks for the purpose of resolving the banking crisis. One-half of the nation's banks with approximately 50 percent of total banking resources were judged to be solvent on March 15, 1933. Of the remaining 50 percent, 45 percent were placed under the direction of "conservators", and 5 percent were closed permanently (Wicker 1996).

Hall and Ferguson

Hall and Ferguson (1996) offer a multivariate explanation of the conditions known as the Great Depression. The overbuilding of housing in the mid-to late 1920's occurred coincidently with the presence of easy credit. A housing slump occurred, followed by an increase in foreclosures, peaking in 1933. Bank failures increased in the 1930-1933 period. Thousands of banks could not meet their depositors' desire to withdraw funds. Banks responded with increased failures. This fear of more failures caused the public to hoard currency.

The early years of the Depression (1929-1933) were severe. High levels of unemployment over the entire course of the Depression were marked by a peak in 1933 (36 percent). It was three years after the Depression when the U.S. economy achieved full unemployment. Despite high levels of unemployment (1934-22 percent, 1935-20 percent, 1936-17 percent) there was increasing inflation in the mid-1930s. Hall and Ferguson (1996) attribute this to programs emphasizing increasing prices and wages that were part of the New Deal.

Bank reserves declined in the 1930-1933 period as depositors withdrew their cash. Hoarding occurred. Depositors shied away from making deposits. This in combination with increased withdrawals resulted in the decline of checking account currency, one of two components of M_2.

According to Hall and Ferguson (1998) the Depression ended due to a rapidly increasing money supply in 1938 and to preparations for World War I which consisted of the U.S. sales of war materials to England and the U.S. defense buildup in 1940.

New Perspectives

What perspectives are presented by this discussion of the Great Depression era?

- The stock market Crash was followed by an increase in the unemployment rate.

- A review of major theoretical views of causes for the Great Depression era is presented.

- The presentation of Stress Theory and its ability to be used in situations of aggregate supply and aggregate demand makes new perspectives on the Great Depression era possible.

- A theoretical disaggregation of the term for personal consumption makes it possible to calculate the level of personal consumption and its variation as affected by variation in mutually exclusive subgroups of the population and their average disposable personal income.

- The lagged immigration effect was highly correlated (r = .84, p = .000, N = 19) with the inverse of the adjusted unemployment rate, although N was small.

- The immigration act passed by the Sixty-Seventh Congress added to this by limiting immigration to 3 percent of the number of foreign-born immigrants registered in the 1910 census. The act was extended to June 30, 1934.

- The increase in the number unemployed (1929-1933) and the decline in the size of the working, foreign-born white immigrant population from 1930 to 1940 as a source of decline in personal income and personal consumption are highlighted. The fact that this had a cumulative impact on personal income and personal consumption makes the effect more significant. The small number of white foreign-born immigrants in the 1931 to 1940 period reveals that there was not adequate immigration (less than 100,000 immigrants per year) of these age groups for replacement to occur.

- The author speculates that this also reduced the number of bank depositors and increased withdrawals from banks.

- A multivariate explanation of the money supply (M_2) with independent variables personal income, the adjusted number of the unemployed, and the wholesale price index has been advanced.

- A multivariate explanation of personal income (PINC) with independent variables M_2, the adjusted number unemployed (ADJUNR), and the wholesale price index (WPI) is offered. The featuring of both personal income and M_2 as dependent variables and independent variables in these two Equations suggest a strong relationship between the two.

- In Chapter XII, checkbook money and central bank money are defined. Central bank money declined from 1930 to 1933. Depositor confidence in banks declined from 1930 to 1933. The decline in confidence as well as the decline in personal income explains the downward trend in bank reserves in 1930-1933.

- The identification of the discouraged worker as a major component of the unemployed population combined with adjustments to the civilian labor force yields a revised perspective on the magnitude of unemployment during the Great Depression.

- The role of the FDIC in insuring bank deposits to a significant level was a major factor in the recovery of banks and perhaps the economy.

- Graphic evidence suggests the possible effects of abandoning the gold standard on a gradual recovery in 1934 and after.

- The theoretical presentation in Chapter X (Equations 10.5 and 10.6) disaggregates the aggregate demand side of the Equation to show how fluctuations in immigrant population numbers and incomes can effect overall personal income, disposable personal income, or personal consumption (especially in 1930-1940).

- Path diagram 2 (Chapter XII) shows the importance of personal income and its effects on bank reserves.

The population decline lesson found later in this book applies to the circumstances from 1961 to 1975. A period rivaling that of the Great Depression in length and severity is predicted primarily due to the decline in effective demand as a result of the peak and decline in live births.

This book represents a movement away from univariate explanations of the Great Depression Era. The new perspectives introduced call for a multivariate view and discussion of the Great Depression's conditions and causes. We shall explore these in the following chapters.

CHAPTER III

AN HISTORICAL OVERVIEW: PRE-1929 TO 1940

The Pre-Crash United States

The citizens and economists of the United States viewed the country as a nation of economic promise. The land of plenty, in the aftermath of World War I (WWI), had seen the arrival of mass production along with the wonders of new technology. Profit-sharing was encouraged by American employers. Much of the stock-buying at the time was on margin. Investors were required to put up 25% of the price of stock. This dubious structure of credit for brokers' loans was halted when in February 1929 the Federal Reserve Board required member banks not to lend money for speculative purposes. This prohibition seemed to go unheard, as some private bankers proceeded to invest large sums with confidence.

By 1929, the buying of goods, such as cars, clothes, washers, furniture, jewelry, and refrigerators, on installment credit had falsely increased the buying power of consumers while helping industry break down sales resistance. By 1929, the ease of installment buying had lured tens of millions of consumers.

In 1870, wage and salary workers made up one-fourth of the United States work force. In 1929, wage and salary workers made up four-fifths of the United States workforce. Agriculture had over-expanded in the period of 1917-1918. This was also true of bituminous coal mining and textiles. In 1926, the unemployed were estimated at 1.5 million. In 1929, the number unemployed increased to 1.8 million. Public relief expenditures gradually increased.

In the six years prior to the Crash of 1929, bank failures (primarily small banks) occurred at an average rate of two per day. At least 20% of the country's productive resources were not being used. This amounted to 25% of the goods and services being produced. Production was not the problem, it was consumption. At the first sign of trouble, consumer goods, whose purchase could be delayed, would back up in warehouses. Joblessness resulted.

From 1923 to 1928, the index of wages advanced from 100 to 112. Over the same time period, the index of speculative gains increased from 100 to 410. This change indicated that the income going into the purchase of consumer goods was small relative to investment, into the call-money market, and into gross domestic private investment.

Income and saving dynamics in the 1923-1929 period tell a significant story. Families with incomes of over $10,000 saved two-thirds of all savings in the United States. Those with incomes of less than $1500 made up 40% of the population. Families making below $1000 per year were 20% of the population. Twenty-four thousand individuals making $100,000 or more earned a total income three times as great as the 6 million poorest families. Mass purchasing power was unable to consume the nation's output. Retail prices remained almost level between 1922 and 1929. The overall output of American labor increased by more than one-third in the ten years following WWI. Yet, savings were not passed on to the consumer.

The Crash

An excellent discussion of the events leading up to and occurring during the stock market Crash (October through November 1929) is given by Mark Smith (2003). He takes a more descriptive view of the U. S. stock market in "A History of the Global Stock Market." A synopsis of that view follows.

Although of little note, the U.S. Stock market reached its high on September 3, 1929. A downtrend continued into early October. On October 24, the decline became a rout. The stock market had turned into a panic in late morning. A crowd gathered outside the market, and the visitor's gallery was closed.

The bankers in New York met and decided to pool their resources to stabilize stock prices. Richard Whitney, president of the New York Stock Exchange, went to the stock exchange floor in the early afternoon and placed bids for a number of major stocks. There were immediate effects. To the satisfaction of investors, the market rose the following day. The psychological impact of the bankers' support for the market had been sufficient to prevent further decline.

Investors remained cautiously optimistic. When Tuesday, October 29, arrived, a shocking 16.3 million shares traded. The decline continued until mid-November, when the stock market and the Dow Jones Industrial Average had declined almost 50% from its September high. A search for individuals or organizations to blame for the stock market Crash occurred in the Crash's aftermath.

Kindelberger's (1989) analysis seemed to be applicable to the events preceding, during, and after the Crash. There was, initially, an ample supply of credit to finance speculation. Easy money fueled speculation as more individuals entered the market. Speculation increased until the stock market peaked and then burst.

The Federal Reserve was said, by some observers, to loosen credit cutting interest rates at a mistaken time, thereby providing credit to the stock-market boom. The easing of credit had been fashioned to halt the flow of gold into the United States during the 1920's. The Federal Reserve had, in effect, eased credit because of foreign exchange considerations.

A new standard of stock valuation had evolved in the 1920s. In place of current dividends, future earnings projections became the new standard in the years immediately prior to the Crash. Friedman and Schwartz (1963) contradict this view, noting that the U.S. consumer price index actually fell over the decade of the 1920s. They also contend that the stock of money actually fell during the expansion period.

Justification of price-earnings (P/E) ratios was confirmed in several studies. They concluded that the stock market did not experience a "financial orgy". Similarly, corporate earnings rose at an annual rate of 9%. Meanwhile, academics and the public retained the view there were financial excesses. Because of these conditions it has been assumed that money was not easy.

The Crash was followed by the Great Depression era.

National Income

National income declined from $81 billion in 1929 to less than $68 billion in 1930, to $53 billion in 1931 and bottomed at $41 billion in 1932. The nation's estimated wealth declined from $365 billion in 1929 to $239 billion in 1932. During the 1929-1932 period, there were 85,000 business failures in addition to the consolidation or failure of approximately 10,797 banks.

The money paid as salaries declined 40%, dividends decreased by 56.6%, and wages declined by 60%. Per capita income declined from an average of $698 in 1929 to $495 in 1932.

Affecting national income was increasing unemployment. In 1930, 3 million were unemployed and 45 million were employed. In January 1931, approximately 8.02 million were unemployed (this book makes an adjustment for discouraged workers in Part II). Four million or 5 million more were expected to be added to the unemployment rolls in 1933.

Exports and Imports

Although much attention will not be focused on the events of the Depression outside the United States, the world at the time was becoming more economically interdependent. Mainly, agricultural lands suffered. Several nations experienced the consequences of depression (Argentina, Australia, Brazil, Canada, Germany, the Near East, and the Orient). After the shock experienced by the United States, Czechoslovakia, France, Great Britain, Scandinavia, and Switzerland felt the effects of the Great Depression.

The Hawley-Smoot Tariff Act as enacted in June, 1930. It further aggravated an already critical situation. The United States had become a creditor country. Private investments had increased from approximately $3 billion prior to WWI to $14 billion in 1932. Foreign dumping of goods in the United States encouraged farmers and industrialists to demand higher protective rates. They obtained the protective rates at an average of 40%. This action was taken as a sign of economic war. By 1932, twenty-five nations had reacted, such that the volume of exports was reduced to half its level before the enactment of Hawley-Smoot. To escape this economic impact, 258 factories had been established in foreign countries.

The growth of cartels was another way in which business protected itself. While domestic cartels remained illegal under federal law, this did not prohibit numerous. American firms from benefiting by the consolidation of their power internationally. Bendix Aviation, Anaconda Copper, DuPont, General Electric, and Standard Oil made agreements with foreign producers in the 1920s and 1930s that often restricted production so that prices could be raised and production restricted. In addition, world markets might be divided and patents exchanged. Foreign and domestic trade was restricted, and the introduction of product improvements and new products was sometimes inhibited. In the Depression, the effect of cartels was to increase under-consumption and increase unemployment.

Isolation and World War I War Debts

American attitudes toward foreign lands were accentuated by WWI debts of foreign nations to the United States. Some saw the debts of Europe to the United States as both hurting the economies of Europe as well as harming the economy of the United States as a creditor nation. When Franklin Roosevelt was inaugurated as president in 1933, most of the WWI debts were in hopeless default. These unpaid debts did much to fuel the pacifism of the American mid-1930s.

The Effect of the Depression on Women

In 1930, one out of five women in the United States worked outside the home, a total of approximately 11 million women. As the Depression proceeded, labor surpluses resulted in low wages and shorter work hours. For example, in 1931 the National Education Association found that three-fourths of all cities had banned the employment of married women. A woman, in whatever field, was assumed to be taking a male's job. Domestic help was often given up as an economizing measure.

In the early thirties, the sale of numerous gadgets which had been made available to the American homemaker declined. Egg beaters, electric toasters, electric stoves, percolators, waffle irons, and washing machines were among those conveniences. Conversely, refrigerators were in increased demand.

Hunger Marches and Hunger

In the spring of 1932, there were hunger marches throughout the nation. The head of Veteran's Affairs reported to Hoover in 1931 that 272,000 men were in need of relief. A cash loan with Adjusted Compensation Certificates was soon provided over Hoover's veto. The Patman Bill provided immediate payment of the balance of the certificates. Later, on June 16, 1932, the U.S. Senate resoundingly rejected the Patman Bill. Veterans who had come to the Capitol to support the bill had to swallow their pride. In 1932, numerous communities were turning over tracts of land to the unemployed for cultivation as gardens or small farms.

Officials repeated assurances that no one would be allowed to starve. Contrary to these exhortations twenty-nine persons in New York City died from starvation in 1933. Philadelphia's community health center experienced an increase of approximately 60% in malnutrition diagnoses between 1928 and 1932. The National Organization for Public Health Nursing reported an increase in malnutrition patients from eighteen percent of admissions in 1928 to 60 percent in 1931.

Unknowing Leadership and Citizens

WWI had convinced citizens that the federal government had a role in crisis. Since that time, government and business had become a part of daily life. Some saw the federal government as balancing the strength of private enterprise. Others thought that the federal government was to be the main actor in times of war, panic, or mass suffering. Many viewed the government as a shield against corporate exploitation and greed.

In the autumn of 1929, Hoover along with many economists and citizens did not correctly estimate the duration or the severity of the crisis. After the Crash in 1929, Hoover repeatedly called financial and industrial leaders to the White House soliciting their cooperation. Major industries pledged to sustain wage rates, Hoover sought further promises of increased spending from the railroads, telephone, and steel companies. Most of the commitments were half-hearted or evasive. Hoover resorted to rhetoric that, in the final analysis, did little to address the situation.

The year of 1930 witnessed some minor efforts to address the Depression. There was aid to farmers to feed their cattle (not their children). Programs were initiated using federal aid to drought victims and a program of public works (including $500 million for federal buildings and $60 million for the construction of Hoover Dam). While such efforts relieved local unemployment, they were not of sufficient scale. Wider action was the need and call of Americans. Hoover attempted an unsuccessful program termed the National Credit Corporation, which he promoted in anticipation that strong banks would voluntarily form a credit pool to help the destitute. Hoover soon became an unwilling scapegoat. The prosperous demonstrated little interest in assisting the poor and the destitute. Thus, the entire burden fell on the government in Washington.

Conservatively, President Hoover disliked the idea of increasing taxes, federal grants (with the exception of loans), and public works that were non-productive such as streets, highways, city halls, state capitols, harbor and river improvements, and extraneous military construction. Alternatively, he favored works productive of income such as toll bridges, toll tunnels, waterworks, and other projects that had the capacity to produce a repayment of the government's investment. In July, 1932, the income-producing projects were favored by the investment of $1.5 billion, while non-productive projects received $300 million.

Housing

There was another significant problem that President Hoover was under pressure to address. The effect of the Depression on the building trades was catastrophic. Between 1928 and 1933, construction of residential housing fell 95%. Repair expenditures declined over the same time period from $500 million to $50 million. Furthermore, in 1932, 273,000 homeowners lost property by foreclosure. Hoover persuaded Congress in July, 1932 to establish twelve Federal Home Loan Banks in order to permit funds to be loaned to building and loan associations, banks, and insurance companies with credit that had been severely strained by loans to residential and farm owners. Mortgage lending institutions were kept solvent, but their assistance to individuals and families proved to be small.

Hoover's Defeat, Roosevelt's Election

In the summer of 1932, Hoover was re-nominated as a candidate for the presidency by the Republican Party. In 1930, Franklin Delano Roosevelt re-entered politics as the governor of New York State. It was a rapid transition for him from the governorship to presidential candidate. The campaigns of the two candidates differed in their views of the Depression. Hoover did not see the necessity of reform. Roosevelt, however, demanded reform as an essential element of recovery. He was concerned about another collapse. He pledged action on old-age insurance, unemployment insurance, control of crop surpluses, credit relief to the states for unemployment relief, and trade agreements with other nations. He promised an effort to develop public works and natural resources for the common interest.

Hoover retained his belief in the efficacy of local self-help for relief and recovery. He proposed the strengthening of the credit structure of business through the use of federal loans. Hoover stood by the gold standard, and favored high tariffs. He waxed eloquent about the retention of the latter fearing what would happen if tariffs were taken away.

Republicans boasted of a mild improvement in the economy in late 1932. The Reconstruction Finance Corporation (RFC) slowed the rate of bank failures, and gold began to return to the United States, the stock market rallied slightly, and the business index went up slightly. For the average citizen, bread lines, soup kitchens, and declining pay, or an increasing unemployment made a change in government leadership more probable. The American public wanted action. Roosevelt's willingness to experiment became apparent as his campaign proceeded. Among those pledged initiatives were federal power plants on the Tennessee and Columbia rivers, protecting the investor against fraudulent claims, easing of farm mortgage burdens, reciprocal tariff agreements, reform of holding companies, and Social Security. Roosevelt pledged fiscal responsibility despite his extensive plans. In particular, Roosevelt attacked Hoover's fiscal responsibility citing the growth of the federal budget from $2 billion in 1927 to $3 billion in 1931.

Soon after his election, Roosevelt declared his overall objective for the American people as a more abundant life. In accepting his nomination, Roosevelt said, "I pledge you, I pledge myself to a new deal for the American people" (Wecter, 1948, p. 53). The New Deal quickly turned into a political tag.

The election was a sweep for the Democrats. Roosevelt was elected by approximately 33 million to 16 million for Hoover. Roosevelt carried all states but six, four of them in New England. The Democrats also took both houses of Congress. Roosevelt was to take his desire to defeat hunger, want, insecurity, poverty, and fear into his first two terms as president.

There were broad theories that were characteristic of the New Deal in its early stages. Keynes (1935), the British economist, was gaining acceptance for his ideas in the 1930s. He argued for compensatory spending in times of depression, a managed currency, and deficit spending. The argument was that production returns should go more and more to consumer and wage-earner while less went to investment and speculation.

Among the foci of the New Deal were a living wage, adequate leisure, economic security for the masses, and restraining the power and great wealth of a few. Those who supported Roosevelt saw this as the only way that the United States could regain its reputation as a land of opportunity. The areas in which private enterprise had failed would now be addressed.

Roosevelt's First Term

Roosevelt was inaugurated on March 4, 1933. During Hoover's lame duck period, Congress continued to take responsibility for the increasing peril in the United States economy. Republicans and Democrats persisted in attacking each other. Democrats blamed Hoover for a fear campaign in October, as commodities and securities declined. Republicans attributed that activity to a concern about Roosevelt's succession. Roosevelt refused to please his predecessor by declining to make joint statements about war debts and fiscal conservatism. Breadlines grew longer while the politicians bickered. The amount of relief available was token, and the nation's banking system was headed for another crisis.

From the beginning of 1930 through 1932, 773 national banks failed. A total of 3,604 state banks met a similar fate. The former had $700 million in deposits. The latter had $2 billion in deposits. Panicking citizens withdrew their deposits and hoarded it by putting the money in safety deposit boxes, tin boxes, canning jars, trunks, or even buried it in their back yard. Circulating money became scarce. In late 1932, some cities in the South printed their own currency. During Hoover's last days in office, his Department of the Treasury estimated that $1.2 billion had been withdrawn from circulation.

By March 1, 1933, ten states had proclaimed banking holidays. On March 4, 1933, the New York banks were closed. The banking system quickly followed suit throughout the nation. Negotiable paper, script, stamps, Mexican and Canadian dollars, and personal IOUs became the media of monetary transactions. Roosevelt's confident inaugural address revealed a man ready for action.

Roosevelt's inaugural speech addressed treating unemployment, assisting in saving homes from foreclosure, and rescuing farms from bankruptcy. He spoke of the need to mend speculation and provide for a sound currency. Americans were taken by the clear shift of government from a defensive posture to one of attack.

The following day (March 5, 1933) Roosevelt called Congress into special session. The next day, Roosevelt halted the export of gold and all dealings in foreign exchange. He called a banking holiday (March 6-9) for the examination of the soundness of banks before their gradual reopening. Title I of the Emergency Banking Act, passed (March 9, 1933) and approved of the action taken by the president amending a wartime measure empowering the president in time of national emergency to regulate or prohibit the payment of deposits by all banks. Furthermore, during the emergency period proclaimed by the president, member banks were restricted from transacting any business unless authorized by the secretary of the treasury with the approval of the president. Additional provisions of the Emergency Banking Act were:

- Title II of the act stipulated the conditions under which certain national banks (those with impaired assets) could be reopened.

- Title III provided for the issuance of non-assessable preferred stock by national banks to be sold to the public or the RFC.

- Title IV provided that emergency issues of Federal Reserve Bank notes be made up to the face value of direct obligations of the United States deposited as security. Alternatively, Federal Reserve Bank notes could be issued for up to 90% of the estimated value of eligible paper and bankers' acceptances acquired under the provisions of the Emergency Banking Act.

- Title IV also provided for Federal Reserve Banks to make available advances to member banks under exceptional circumstances.

After the emergency declared by the presidential proclamation had ended, Federal Reserve Bank notes could be issued only using the security of direct obligations of the United States.

The RFC provided new capital to issue more currency and supply new capital to reorganize banks. The banks were unfrozen to allow for essentials such as medicine, pay rolls, and relief funds. An executive order on March 10, 1933, gave the secretary of the treasury power to grant licenses to member banks to reopen. Member banks were to make application to the Federal Reserve Bank of its district, which would act as an agent of the secretary of the treasury in granting licenses. The executive order also provided for state banks to reopen sound banks that were not members of the Federal Reserve System. A later executive order (March 18, 1933) provided state banking authorities with the authority to appoint conservators for unlicensed banks when not in disagreement with state law. A program for reopening licensed member banks as well as those licensed by state banking authorities took place on March 13, 14, and 15. On June 16, 1933, The Glass-Steagall Act was passed and initiated several banking reforms. It provided for deposit insurance, separated commercial from investment banking, and gave the Federal Reserve power to prevent loans for speculation.

In March 1933, the Securities Act was passed, followed by the Securities Exchange Act in 1934 and the Public Utility Holding Act in 1935. This complex of legislation put a limit on bank credit for speculative purposes, established safeguards against stock manipulation, provided for full information for those purchasing securities, created the Securities and Exchange Commission to oversee the stock market, and ended most utility holding companies with specific exceptions. More controversial were the New Deal's monetary measures. Increasing agricultural prices relative to nonagricultural prices was a priority. Because wheat and cotton were subject to international demand, it was thought that devaluing the dollar would relieve the situation. The government, following Great Britain's example, went off the gold standard in April 1933. In June, Roosevelt received permission from Congress to inflate the U.S. currency in five defined ways. Roosevelt, along with France and other gold bloc countries, undercut the world monetary and economic conference in June, 1933. Other nations became hostile to the United States while in the United States the average citizen's isolationism was confirmed. It was not until September 1936 that Britain, France, and the United States finally reached agreement on preventing large fluctuations in the value of their currencies and adopted measures to prevent devaluation.

Devaluation of the dollar in late 1933 forced the dollar to less than 60% of its former gold content. This was expected to improve the position of American exports and to increase prices generally. A marginal improvement in foreign trade occurred. The primary visible effect was the flow of gold from other nations into the vaults at Fort Knox. The gold purchase plan operated from October 25, 1933, to June 19, 1934. It failed to appreciably increase domestic commodity price levels while its day-to-day variation disrupted confidence and stability.

The Silver Purchase Act was forced through Congress on June 19, 1934, by the silver block with the agreement of the president. Although its intent was inflationary, it resulted in both boosting the price of silver and the acquisition of $1 billion dollars of foreign silver at prices above market value. The effect of these purchases was not the moderate re-inflation which the planners had intended, nor the rapid inflation which Wall Street observers had foretold. This law could be put down as one of the mistakes of the New Deal.

It was assumed that money and credit, and the price and profit system had broken down because of the Depression. The Roosevelt administration knew that to let the economy take its course would be slow and likely detrimental. The use of government spending to stimulate the economy and ease the way towards prosperity for the workers and consumers was seen as a primary tenet. Producers were consumers and vice versa.

It was argued that five related processes would benefit the entire nation. These included: the creation of government-financed work; developing industrial codes to increase employment and wage rates; increasing farm income by restricting crop plantings; making direct farm payments; and raising the general price level of goods and services by currency regulation with federal support and credit institution regulation.

Roosevelt warned in his inaugural address that the federal government was headed for bankruptcy. This was followed by the congressional passage of legislation to reduce federal payrolls and reduce veterans' compensation by approximately $400 million that year. The American Legion lobby and traditional American politics defiantly opposed the cuts. Roosevelt used his powers of executive order to gradually restore the cuts. The president soon led the spending parade. He requested billions for relief, more for the costs of running new federal agencies, and dollars for pump-priming. In March 1934, any traces of gestures on the part of the President to be economical were vetoed by Congress.

March, 1933 witnessed the return of 3.2 beer. Hoover and Roosevelt differed on the repeal of Prohibition as the Republicans hedged their bets by being opposed to Prohibition's repeal. Farmers needed to sell their grain and sugar, and Democrats were emphatic in their support for repeal. The Twenty-First amendment achieved the required backing (thirty-six states) in December 1933, and the U.S. social experiment with Prohibition ceased.

In the spring of 1933, the banking crisis was easing, and the attention of the Roosevelt administration focused on relief. There were approximately 13 million persons unemployed and about 6 million receiving state and municipal charity. On March 21, 1933, the president delivered a message to Congress calling for three types of remedial legislation. These included: grants to states to provide direct relief for feeding and clothing the destitute, a program of lasting public works, and the enrollment of workers in programs that could be quickly established and would not interfere with private industry.

The Federal Emergency Relief Administration (FERA) was established by Congress on May 12, 1933, to help the needy with money for direct relief. It was administered through existing channels of state and municipal welfare. The second and third of the preceding criteria contained the origins of the Public Works Administration (PWA) and the Civilian Conservation Corps (CCC) respectively. The Federal government supplied over seventy cents of the relief dollar. The states supplied thirteen percent of each relief dollar and municipalities supplied approximately sixteen percent.

Soon, make-work projects at minimum wages grew into favor. They were praised as they raised the worker's self-respect. Unions were concerned about the effect of work project wages on private industry. Some individuals supported relief as a cheaper alternative that would decrease the indebtedness of the federal treasury. A large majority of people endorsed the choice of work projects. A Gallup poll taken in May 1937 revealed that 80% those polled approved of work relief by public works.

FERA, in the summer of 1933, included work relief tasks such as counting vehicles at intersections, picking up papers in parks and along roadsides, and raking leaves. Care was taken to ensure that pay levels did not interfere with those of private industry. A minimum wage of thirty cents an hour was established with the exception of the South and a few industries. A newly reconstituted U.S. Employment Service also made jobs available to the unemployed.

In January 1935, approximately 2.5 million workers were employed by FERA. The 1933-1934 winter witnessed FERA's widest expansion. There were 28 million persons, or 8 million households, on relief in February 1934. This marked expansion occurred because of the addition of a special work-relief organization termed the Civilian Works Administration (CWA) in November 1933. Approximately one-half of the individuals came from the regular FERA roles. The other one-half were taken from the non-relief unemployed. CWA devoted itself to a wide variety of projects. Among CWA's projects were pest and erosion control; work on municipally-owned utilities; repair of schoolhouses, parks, and playgrounds; and repair of roads. The CWA spent approximately $900 million in total. In the spring of 1934, the CWA was terminated with unfinished projects being turned over to the FERA program. December 1935 saw FERA's demise.

The effects of the PWA were more robust. Organized in June 1933, the PWA had a budget of $3.3 billion. The PWA was involved in public works that required large volumes of material. It was designed to assist heavy industry. Much of the work of the PWA was done under contract with private businesses. Projects were carefully reviewed by PWA before funding one-third to one-half of their cost. Other projects the PWA undertook independently. In 1938, PWA devoted its efforts to these solely.

PWA was thought by New Dealers to be their best hope of increasing private employment. In addition to having a stimulating effect on industry, it kept 500,000 men working thirty-hour weeks through 1934. Private investments had beneficial effects on local and regional economies. PWA funds brought Hoover Dam into existence two and one-half years ahead of schedule. Other projects included the Triborough Bridge (New York), the Tennessee Valley Authority, farm-to-market roads, hospital beds, model housing, school bus roads, school and university buildings, sewage systems, slum clearance, and water supply works. These efforts included building fifty military airports and 74,000 miles of highways among other projects with a strategic purpose. Of particular note is that the number of federal employees increased from 588,000 civilians in 1931 to 1.37 million in 1941.

In the early summer of 1933, The National Recovery Act (NRA) was implemented. Its purposes were to abolish child labor, get unemployed men back on the payroll, improve consumption, put a floor under prices, raise wages, reduce competitive wastefulness, shorten work hours, and strengthen collective bargaining. Representatives of approximately 800 industries came to get codes of their own. About 2 million workers out of approximately 13 million unemployed were rehired. The average citizen became confident that something was being done by the government. Administrative ineptness, cries of unfair competition by business, and rising prices that outstripped wages were among initial warning signs. Unfortunately, the NRA, by its actions fixing wages on a man-hour basis rather than by production units, affected the smaller and less mechanized industries. Policing by the federal government became ill-financed and half-hearted. Code revision and enforcement had become so cumbersome that Congress debated the legislation's extension beyond its initial two years. The Supreme Court made any action by the Congress unnecessary by invalidating the NRA. In the end, Roosevelt had pinned too much faith on the unselfish and enlightened participation of business.

NRA's best features were revived by the Robinson-Patman Act of 1937 and a series of laws beneficial to labor passed between 1935 and 1938. Despite the increasing fears of business about intervention, the caprice of bureaucracy, confusions and inconsistencies of the NRA, the Agricultural Adjustment Act (AAA), the stock-exchange bill, labor policies, and the fear of communist infiltration of the government and autocratic power, the Roosevelt administration continued its course of action.

The first New Deal budget for fiscal year 1934 (FY) was 2 billion dollars larger than that for FY 1933. Approximately 60% of all dollars went to recovery and relief.

Before the close of FY 1936, the government found itself with $30 billion of debt despite higher taxes. The National Association of Manufacturers (NAM) in 1934 opposed legislation to increase taxes on gift and inheritance taxes, raising taxes on incomes over $50,000 and passing a graduated corporate income tax. NAM saw these taxes as an abuse of federal power that discouraged success and thrift. The Revenue Act of 1936 furthered strengthened the opposition by enacting an undistributed profits tax. The yearly cost for all levels of government rose from $11 billion in 1929 to $17 billion in 1938.

The sales tax, introduced by West Virginia in 1921, was adopted by twenty-one states from 1930 to 1935. Some municipalities also adopted the tax. While the sales tax weighed more heavily on the poor than on the wealthy, it was favored by chambers of commerce and bankers' associations who wished it to be adopted as a federal policy. Congress and the Roosevelt administration gave the sales tax little support.

The repeal of Prohibition at the national level represented a source of revenue. Eight states, five of them southern, opted to remain dry. Fifteen other states chose to make the selling of liquor a state monopoly, with seven of the fifteen allowing for private sale of liquor under special conditions. Local-option laws became more popular towards the end of the 1930s. The taxes on liquor were high, thus continuing to encourage bootleg trafficking.

There was a strengthening of workmen's accident compensation laws by the states in the 1930s. In 1937, thirty-eight states revised and enhanced their laws. The revisions included expanding accident and death benefits, increasing coverage, and improving the definition of occupational diseases. Other provisions that were revised in favor of the worker were dental, hospital, nursing, prosthetic, and rehabilitation benefits. A payroll tax was initiated to support state Social Security systems.

The year of 1935 proved to be a pivotal year for the Roosevelt administration. NRA's collapse and criticism from a variety of sources came as elections were forthcoming in the fall of 1936. Passage of significant pieces of social legislation occurred in 1935. This legislation included the National Labor Relations Act (NLRA), the Public Utility Act (PUA), the Social Security Act (SSA), the Wealth Tax Act (WTA), and the Works Progress Administration (WPA) Act. The WPA separated work relief from direct relief. Direct relief was handed back to the back to states and localities. In 1939-1940, about 60 percent of the expenditures came from state funds.

Monthly wages for employable individuals in the WPA averaged between fifty and sixty dollars a month. The WPA made every effort to devote its funding to wages. Local tax-supported government bodies assisted WPA projects by providing tools and materials. This contribution was estimated at 25 percent of costs. The state and local governments implemented the final products upon their completion. Poorer states and municipalities did not fare well because of their inability to meet the matching requirements.

Projects that were being passed up by businesses and public initiative were addressed by the WPA. The list of accomplishments was impressive. These accomplishments included: auditoriums, courthouses, drainage ditches, hospitals, mosquito control, the building or rebuilding of 110,000 public libraries, half a million sewage connections, 1 million new privies, and water purification. The bulk of the women in WPA were organized into sewing groups that made over 300 million garments for needy adults and children. These groups totaled 30 million persons at their maximum strength. In addition to supplying approximately 600 million school lunches, the WPA organized and maintained 460 nursery schools. For schools and public libraries, approximately 80 million books were repaired. The list did not stop there. Art, assembly halls, dental and medical examinations, immunizations for smallpox and diphtheria, naturalization classes, nursery and other classes, and public golf courses were among the additional projects.

Employment and Unemployment

An estimated 13 million were unemployed in March 1933. Unemployment was a difficult problem for the New Deal and perhaps its most difficult one. Table 3.1 provides a characterization of employment and unemployment from 1925 to 1940.

The unemployment rate was at a low of 3.2% in 1929. This was the low for the 1929-1940 period. The unemployment rate increased to 24.9% in 1933. The unemployment rate decreased to 14.3% in 1937, increased to 19% in 1938, and declined to 14.6% in 1940. The labor force (column 4, Table 3.1) increased at a regular rate from 1925 to 1940. This supposed regularity was disturbed by the changing levels of unemployment (Column 5, Table 3.1) from 1929-1940. These levels were well above unemployment levels from 1925-1928. Later, in Chapter XI, the unemployment data will be adjusted for discouraged workers.

Roosevelt's Second Term

Employees of WPA staff in the 1937 Presidential election aroused unfavorable publicity by campaigning. WPA staff in Kentucky, Pennsylvania, and Tennessee were involved.

Table 3.1: Labor Market Statistics: 1925-1940*

Year	Employment	Farm Employment	Non-farm Employment	Total Labor Force	Number Unemployed
1925	43,716,000	10,662,000	33,054,000	45,431,000	1,453,000
1926	44,828,000	10,690.000	34,138,000	45,885,000	1,801,000
1927	44,856,000	10,529,000	34,327,000	46,634,000	1,519,000
1928	45,123,000	10,497,000	34,626,000	47,367,000	1,982,000
1929	46,207,000	10,541,000	35,666,000	48,017,000	1,550,000
1930	44,183,000	10,340,000	33,843,000	48,783,000	4,340,000
1931	41,305,000	10,240,000	31,065,000	49,585,000	8,020,000
1932	38,038,000	10,120,000	27,918,000	50,348,000	12,060,000
1933	38,052,000	10,090,000	27,962,000	51,132,000	12,830,000
1934	40,310,000	9,990,000	30,320,000	51,910,000	11,340,000
1935	41,673,000	10,110,000	31,563,000	52,553,000	10,610,000
1936	43,989,000	10,090,000	33,899,000	53,319,000	9,030,000
1937	46,068,000	10,000,000	36,068,000	54,088,000	7,700,000
1938	44,142,000	9,840,000	34,302,000	54,872,000	10,390,000
1939	45,738,000	9,710,000	36,028,000	55,588,000	9,480,000
1940	47,520,000	9,540,000	37,980,000	56,180,000	8,120,000

* Source: Historical Statistics of the United States, Part 1, p. 126, col. 5, 6, 7, 1, 8

The development led to the Hatch Act in 1939, which was designed to restrict political activities by federal employees.

Construction outlays increased from an annual average of $188 million between 1925 and 1929. Between 1933 and 1938 annual average construction expenditures were $1.63 billion. Nonetheless, the National Resources Planning Board reported that the effect of federal public works programs was negligible.

The Democrats ran the 1936 election on the progress Roosevelt had made with the New Deal. The Republicans, assembling for their convention in Cleveland, met in an anti-Roosevelt atmosphere. In a genuflection to the program of the New Deal, the Republicans recognized society's obligation to provide for the security of people by protecting them against voluntary unemployment and dependency in old age. Otherwise, Republicans promoted payments to farmers for soil conservation, the retirement of non-productive land, and federal stimulation of cooperative marketing. Labor was promised collective bargaining without interference. The Republican platform favored state laws to punish child labor and sweatshops. This was done in the face of the Supreme Court's ruling that a New York minimum wage law for women infringed on the "freedom of contract" between employees and employers. This decision appeared to close the door on all state regulation of working conditions.

Roosevelt was not intimidated. He took the Republicans head-on, declaring that the Republicans wished to restore the United States to their governance, which is, at most, indifferent. After the election, an analysis of campaign funds found that the Republicans had approximately $9 million to the Democrats' $5.25 million. Roosevelt received 61% of the popular vote. This compared favorably to Roosevelt's 57 percent victory in 1932. Roosevelt took forty-six states.

Encouraged by the voters' response to his re-election campaign, Roosevelt, in 1937, initiated a proposal to reorganize the Supreme Court. His approach required that judges retire from the bench at age seventy. If they did not, an additional judge would be appointed up to a total membership of fifteen. This effort would have an effect desired by Roosevelt, as only one of the six judges seventy or over was of liberal persuasion. In the spring of 1937, Justice O. J. Owens shifted to the liberal bloc. Chief Justice Hughes seemed to vote increasingly for liberal issues. The Court upheld the Railway Labor Act, upheld the altered Frazier-Lemke farm mortgage moratorium, and sustained the State of Washington's minimum-wage law. Additionally, the Wagner Labor Relations Act and the unemployment-insurance tax provisions of the Social Security Act were upheld. There was a consistent five to four vote on each of these cases. While the Senate rejected Roosevelt's Supreme Court proposal in August, 1937, Roosevelt attained his ends.

Defining the Interests of Unions and Management

While the actions of Roosevelt served to set the businesses and businesspersons against him, labor found important reasons for supporting him. The year of 1929 followed a decade after WWI in which management was firmly in control, and unions were declining in their numbers. Southern textile mills, steel companies, and automobile companies were not penetrated by unionism. It was at this time that mass unemployment occurred. This development was accompanied by declining numbers of dues-payers and a decline in collective bargaining.

However, labor both won and lost important legal victories from the Hoover administration. Company-sponsored unions were a thorn in the side of organized labor. The Supreme Court decided in the case of *Texas & New Orleans Railway Co. v. Railway Clerks* (1930) that workers' rights were interfered with by an employer's effort to force a company union on employees. Furthermore, in March, 1932, Hoover signed a law that bound employees to sign contracts that forbade them to join unions. In the same legislation, the federal courts were forbidden from issuing injunctions against unions.

Section 7A of the NLRA emphasized phrases of the Norris-LaGuardia Act by guaranteeing employees the right to organize and bargain collectively through representatives of their own choosing with no restraint. Almost immediately, management and labor sought to set this guarantee to contrary purposes. Management made efforts to increase the rate at which company unions were formed. Labor attempted to destroy company unions and make its own unions exclusive bargaining units. Membership in the United Mine Workers (UMW) increased from 150,000 in 1932 to 400,000 in 1935. The membership in the American

Federation of Labor (AFL) increased approximately 75% between midsummer 1933 and midsummer 1936.

A floor under wages and a limit on hours were mandatory for employers who signed on to the NRA. This practice was occasionally evaded; meanwhile employees were dissatisfied with the lag between increasing prices and increasing wages. The restriction on hours (reiterated by the Wagner Act) and depressed conditions resulted in a significant shortening of the work week. A five-day work week became the rule.

The president created the National Labor Board (NLB) composed of employer and union representatives in August 1933. Its powers of enforcement were almost nonexistent. For this reason and Congress' interest in reasserting itself, Congress substituted the National Labor Relations Board (NLRB). The NLRB had quasi-judicial functions and was composed of three labor-relations specialists. Although the NLRB was supposed to be impartial, business management soon started complaining when the NLRB often sided with employees. Among the decisions of the NLRB, the most opposed by the NAM was one in which the NLRB held that a business could not bargain with minority interests, but must be a sole collective agent representing a majority. Employees were not compelled to join such organizations.

While the NRA and the NLRB collapsed, labor was powerful enough to lobby Congress for additional legislation to maintain its gains. Labor had become more vociferous and demanded from the government bolder tactics. The Wagner Labor Relations Act was passed in July 1935. It did not allow business interference with collective bargaining and the right to organize. Furthermore, it prohibited refusing to deal with employee representatives, the development and promotion of company unions, and did not allow discrimination in employment. Again a three-member NLRB was appointed. Some states followed suit by passing legislation with similar prohibitions applying to intrastate commerce.

By the end of January 1941, the NLRB was involved in more than 33,000 cases concerning 7 million employees. Of those cases, 3,166 were strike cases involving 400,000 employees. In all, 2,383 strike cases were resolved. Many of the remainder, involving 200,000 employees, averted strikes. The mid-1930s witnessed an increasing number of strikes, as workers became bolder in airing their grievances. In some instances, business responded. For example, under the threat of unionization, General Motors from January 1934 to July, 1936 expended approximately $1 million on private detectives to infiltrate strike breakers.

The threat to automobile makers came from a new force in American labor, the Committee for Industrial Organization. John L. Lewis was one of the more aggressive leaders spearheading this movement. In November 1935, Lewis and other union leaders who wanted to form industrial unions formed the Committee for Industrial Organization inside the AFL. Initially, the Committee for Industrial Organization formed the United Automobile Workers (UAW) in August, 1935, followed by the United Rubber Workers (URW) in December. The executive council of the parent union (AFL) ordered the dissidents to disband. In August 1936, the maverick unions were suspended. The industrial unions shortly seceded. In 1938, the membership of the industrial unions was 4 million. It topped the membership of the AFL by approximately 500,000. These more radical unions called themselves the Congress of Industrial Organizations (CIO). These industrial unions favored fighting tactics, such as using the strike more frequently to obtain their ends. The CIO also supported a legislative program to the left of the New Deal's. Labor, led by the CIO, became a force for political action. The New Deal, aware of its commitment to pro-labor policies and the support of organized labor, was reluctant to restrain the unions. This was a significant change, as big business had previously received this preferred treatment.

The year 1937 was a record one for strikes. There were 4,720 strikes in total. Many of them (2,728) were waged over the right of unions to organize. The sit-down strike had become a popular tool of unionists. Soon, because of popular and government opposition, unions recognized its limited usefulness. In some ways, the frequency of strikes was a sign of the growing strength of the labor movement. The government was affected by efforts to organize relief workers and the Project Workers Union (PWU). The latter applied strong pressure against WPA wage cuts and layoffs. This development occurred despite the president's denial of the union's right to strike.

In late 1937, the recession was blamed by many in the conservative press on labor turmoil and work stoppages. The accompanying unemployment weakened the bargaining power of labor. In 1938, there were only 2,770 strikes. This was approximately one-third the number of the previous year. The strikes involved approximately 690,000 employees. It was paradoxical that the president and his staff carried through Congress the Fair Labor Standards Act (FLSA) in June, 1938. The FLSA extended to all work affecting interstate commerce. Only farming, fishing, and some types of selling and service were exempt. A maximum work week of forty-four hours was established (to be reduced to forty hours), the abolishment of child labor in the preparation of goods for interstate commerce, time-and-a-half pay for overtime (except in some seasonal occupations), and a minimum wage of twenty-five cents an hour that was to be gradually increased to forty cents per hour. Immediately, the effect raised hourly pay for 300,000 workers. The work week was shortened for 1.3 million workers. The following year's wage increase benefited about 690,000 people. The decrease of the work week to forty-four hours benefited about 2.328 million employees, and movement to the forty hour work week in 1940 affected 2 million workers.

Post-1939 defense and war needs increased employment. Gradually, the reserve pool of unemployed manpower was drawn down. Factory workers became a more satisfied lot as their real weekly earnings in 1941 were at an all-time high. Strikes gradually decreased.

A sign of the new functions of government occurred with the passage of a law reorganizing the federal government in 1939. A single Federal Works Agency was formed that included welfare activities such as those of the PWA, WPA, and U.S. Housing Authority. This was an effort to reorganize a number of the New Deal's creations. The federal government had become a force in the life of everyday citizens.

Urban and Rural Changes

In 1935, urban America's ten biggest cities contained one out of five of the employables on relief. Many homeowners were stressed, as they did not have the funds to keep up with their home mortgages. In June

1933, the New Deal established the Home Owners Loan Corporation (HOLC). Loans were refinanced at 5% interest and were amortized over a fifteen-year period. Home repairs were made under supervision, and their costs added to the cost of the loan. The lending period ended in June 1936. At that time, the HOLC had supplied over one million loans to homeowners costing the government over $3 billion initially. Approximately one-sixth of the urban home mortgage debt in the United States had been assumed. The number of foreclosures, which had quadrupled from 1926 to 1933, had declined by half the number in 1933. Many families were saved from eviction. In 1933, the PWA established an Emergency Housing Division (EHD) to issue private contracts for the clearance of slum areas. The EHD substituted the building of decent homes in place of the slums. Projects were undertaken in Atlanta, Brooklyn, Chicago, and Cleveland. All did not go well, for in New Orleans two ventures were dropped by the PWA when the Huey Long-controlled legislature inundated the projects with local politics. Rentals averaged twenty-six dollars a month. If a resident's family income rose to $130 or $156 a month it was required to leave. When this program was terminated in November 1937, it had sponsored fifty developments with approximately 22,000 dwelling units.

In June 1934, the National Housing Act established the Federal Housing Authority (FHA). It was designed to increase private building with federal mortgage insurance and assist owners of existing dwellings with financing the repair, renovation, and enlargement of existing homes. The Act provided loans at reasonable interest rates, required that standards of construction be met, and provided expert engineering and architectural advice. From 1934 to the end of 1940, FHA had underwritten $1.25 billion to modernize 3 million housing units, provided $3 million for the construction of 600,000 small homes, and funded over 300 rental projects.

In August 1937, the Wagner-Steagall Act was passed. It set up the United States Housing Authority (USHA) to attack the problem of low-cost housing. The USHA was housed in the Department of the Interior and was charged with lending, or less commonly, giving $500 million (increased to $800 million) to local housing agencies for slum clearance, new construction, and repairs under federal supervision. Local participation took the form of often exempting the properties from real estate taxes.

Rent was planned for families with annual income not exceeding $1,000 dollars. In cities over 500 thousand in population the income ceiling reached $1,250 dollars. In some areas, rents were adjusted to incomes as low as $600 in the North and $300 in the South. Almost one-third of the units federally financed in this way were for the use of African-Americans. By 1941, approximately 200,000 family units had been constructed and rented. Despite this fact, criticism remained that such housing benefited the lower middle class more than it did the poor.

In 1939, residential construction units passed the 1 billion mark for the first time since the beginning of the Depression. Despite this fact an estimate was made that the United States remained 4 million housing units short. Because of defense needs in 1941, housing construction drew to a halt. The country's building resources became dedicated to the war effort.

Farms and Farming

In 1937, the Resettlement Administration was absorbed into the Farm Security Administration (FSA). In a study of rural problems, from one-half to three-fourths of all relief families in Southern towns lived in housing unfit for human habitation. Many were faring hardly better. Owners of eroded land, farmers of parched acres on the Great Plains, and sharecroppers suffered. Farm-less, homeless, and jobless families sought low rents, low taxes, and the ability to apply for relief. These persons migrated in increasing numbers to the growing agricultural villages. The idle factories in the cities, the farm tenant class in the South, and drought victims in the Midwest produced migration to the Middle Atlantic states.

The objective of resettlement was to give people the chance to move from bad situations to good. Self-support was to replace dependency and discouragement. Destitute and stranded rural families could purchase, with long-term federal loans, farms scattered in good agricultural regions. The federal government purchased large tracts of land to be subdivided among individuals for lease or purchase. This act often included development of a network of public amenities such as roads, schools, and water supplies. The land many such farmers left was purchased by the government and turned into forests, Indian reservations, pasture, or wildlife sanctuaries. Such projects provided homes for approximately 10,000 thousand families.

Unfortunately, the difficulties of the farm population were multiple. In addition to poor soil, scarce credit, overwhelming debts, drought and flood, ignorance, and pests were other burdens b burdening the farmer. Under the economic and related farm difficulties, mechanization and farm consolidation continued. In Figure 3.1, the increasing number of acres per farm employee is highlighted.

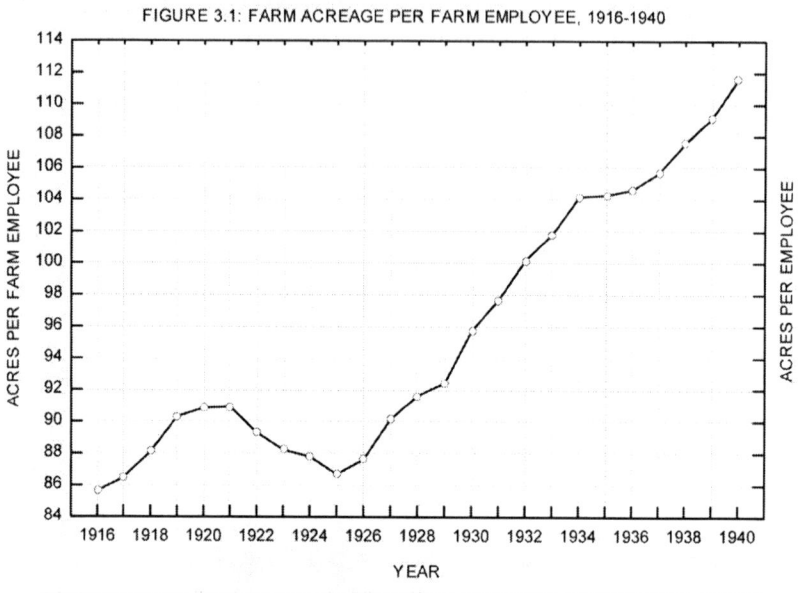

FIGURE 3.1: FARM ACREAGE PER FARM EMPLOYEE, 1916-1940

SOURCE: CALCULATIONS FROM THE HISTORICAL STATISTICS OF THE UNITED STATES

Farm acreage, Part 1, p. 457, col. 5; Farm employees, Part 1, p. 467-468, col. 174

Several observations can be made about Figure 3.1. Progressing from left to right, the period from 1921 to 1926 represented a difficult period for agriculture. World agricultural prices and farm productivity suffered. From 1926 on, four processes went on simultaneously. First, there was increasing mechanization of farms so that the number of farm employees required for a given amount of acreage to be managed declined. Second, the requirement for less labor meant a migration into cities and agricultural villages. Third, agricultural prices remained low. The relative size of the non-farm and farm populations changed dramatically. Fourth, consolidation of farms took place. Figure 3.1 demonstrates the resulting increase in number of acres per farm employee, Figure 3.2 shows the trends in farm and non-farm employment.

Non-farm employment increased from 1921 through 1929. It increased from 26.618 million persons to 35.666 million persons, or 34%. From 1929 through 1933, the number of persons employed declined from 35.666 to 27.962 million, or 22%. From 1933 through 1940 employment increased from 27.962 million persons to 37.980 million. This was an increase of 10.02, million or 36%. A period of decline occurred in 1938.

FIGURE 3.2: U.S. NONAGRICULTURAL EMPLOYMENT AND FARM EMPLOYMENT

SOURCE: HISTORICAL STATISTICS OF THE UNITED STATES

Source: Non-farm employment, p. 120, col. 7; Farm employment, p. 120, col. 6

Farm employment declined from approximately 10.8 million in 1916 to 9.55 million in 1940. This represented a decrease of 1.25 million persons over a 25-year period. From 1929 to 1940, farm employment declined from 10.6 million to 9.55 million. This twelve-year period saw a decline of 1.05 million employees. This was a pace of decline of 87,500 persons per year. This decline occurred against a background of varying surpluses and poor farming years. One in four rural households (3.5 million) required some form of private or public relief. Relief was harder to get, the family allotment was smaller than those in urban areas, and residence requirements were difficult to comply with.

Sharecroppers belonged to the bottom rung of landlord-tenant relationships. Cash-renting, in which the tenant supplied the working capital, paid a fixed rent, and kept the profits, was among the most desirable of landlord-tenant relationships. In the South, this relationship was virtually unknown. Another approach was crop-share renting. The landlord met some production expenses, while the tenant provided the labor, work animals, seed, and tools. A portion of the crops was paid in the form of rent. Dairy and cattle-raising areas of the Midwest used similar sharing of assets and profits. In the South, sharecropping dominated. The tenant traditionally provided his own and his family's labor in return for one-half the cotton and one-third of the grain raised. In some cases, the sharecropper borrowed from the landlord to purchase food and clothing. In other situations, the sharecropper would turn to a credit merchant who would charge usurious rates on loans.

The following table (Table 3.2) provides an overview of decennial data on farms and farm ownership. (This Table will be repeated and added to in Chapter VIII, Table 8.2)

Table 3.2: Farms and the Types of Farm Ownership*

YEAR	FARMS	FULL-OWNER	PART-OWNER	MANAGER	TENANT
1900	5,740,000	3,202,643	451,515	59,213	2,026,286
1910	6362,000	3,355,731	593,954	58,353	2,357,784
1920	6,454,000	3,368,146	558,708	68,583	2,458.554
1930	6,295,000	2,913,052	657,109	56,131	2,668,811
1940	6,102,000	3,085,491	615,502	36,501	2,364,923

* Historical Statistics o the United States, Vol. I, p. 465, Series K 109-153

The number of farms declined between the decennial years of 1920, 1930, and 1940. The decline in full owners between 1920 and 1930 is one indication of the difficulty farms were having in that decade. The beginning of the consolidation process, and the increasing mechanization that was occurring resulted in part owners decreased during the 1910 to 1920 period. This was followed by an increase in the 1920 to 1930 period. Managers decreased in the 1920 through 1930 period, tenant farms actually increased and then decreased from 1930 to 1940. Both indicate the stress on farmers in those decades.

The decline of farms, coupled with the growth in full-time owners, was indicative of the consolidation in farms, farm foreclosures, and the increasing mechanization in the 1930 to 1940 period. The decline in tenancy from 1930 to 1940 is due, in part, to the efforts of the FSA.

Several state legislators passed mortgage moratorium laws, thus anticipating the federal Frazier-Lemke Act passed by Congress in June, 1934. While this Act was later annulled by the Supreme Court, a more moderate version was soon passed by Congress and signed into law. This law delayed mortgage foreclosures for six years, if the affected farmers paid a rent fixed by a federal district judge.

The FSA was established by the Bankhead-Jones Act on July 22, 1937. The initial appropriation of $10 million was increased to $25 million in the next year and $40 million in the following year. It was designed to make low interest (3%) available for forty years to sharecroppers, farm laborers, and other farm tenants to purchase their own farmsteads. The FSA also provided small loans for funds to adjust debt with creditors. These funds went to set up cooperatives, for periods of crop failure, drought, and flooding. The rural cooperatives provided needed equipment and services, including medical care. Important services were supplied by the FSA county supervisor and his staff. The staff of the county FSA taught lessons in thrift management, the use of pressure cookers, and scientific methods of canning. The FSA sponsored the management of 161 experimental and demonstration homestead projects. Some of these projects were run by farm cooperatives.

The Resettlement Administration and its replacement (the FSA) gave financial aid to 1.25 million families between 1935 and 1939. This aid and assistance was designed to move families towards self-support. The decline in the number of tenants and the increase in full owners in the 1930 to 1940 period were in part a result of the Farm Security Act.

The independent farmer had been suffering from depressed conditions from the years after WWI through the 1930s. The farmer's share of national income fell from 15% to 7% in 1933. There were multiple causes. Among the causes were surpluses on world markets, the Hawley-Smoot tariff of 1930, low prices for crops, high prices for the goods necessary for production, droughts, soil depletion, and the lack of capital for mechanization. For cotton farmers, there was cheaper production internationally, and the rising popularity of synthetic goods.

Some farmers prospered. The per-capita consumption of citrus doubled from 1920 to 1940. Milk farmers and middlemen produced and sold $1.355 million of product. The truck farmer benefited by the doubling of the use of vegetables over a fifty-year period. Nonetheless, through good seasons and bad, farmers had reason to complain about low price levels.

In June 1929, Hoover's passage of an agricultural marketing act was responsible for establishing the Federal Farm Board (FFB). The act set up farmers' cooperatives and corporations for financing the control and purchase of farm surpluses. The FFB desired to create steady farm prices and eliminate the excessive profits of commission men. Two factors interfered. The coming of world depression and the farmers' ignoring of voluntary production controls were unaccounted for. Yet, the FFB had no way of controlling such events.

In 1931, the South experienced a large cotton crop. The FFB, in desperation, proposed that every third row of planted cotton be plowed under. By the summer of 1932, cotton was selling for under five cents a pound, and corn at thirty-one cents a bushel. Realizing the scope and severity of the problem, the FFB unsuccessfully called for legislation to allow the federal government to control farm output. After two years, the FFB offered its surpluses to the Red Cross. The New Deal was to learn by these experiences.

Revolt became commonplace among farmers. In Iowa, a farmers' holiday was declared so that prices could recover. Roads were blocked, and milk cans were emptied into ditches. Similar demonstrations were carried out in parts of the Midwest, the South, and East. Prices did not increase, and the demonstrations soon abated. In May 1933, with the farmers' desire for higher prices unrelieved, farmers voted for a national strike. The strike never occurred, as the farmers chose to give the New Deal a chance.

The president consolidated all credit agencies on March 27, 1933. Agencies involved were the FFB, the Farm Loan Board, and some functions of the RFC. These agencies or parts of agencies formed the Farm Credit Administration (FCA). The purpose of the FCA was to assist farmers with their debt by reducing their mortgages and interest payments. For the year ending in mid-March of 1936, mortgage foreclosures declined from thirty-nine per thousand in the spring of 1933 to twenty per thousand.

The AAA, enacted on March 12, 1933, was an important step for the farm economy. To increase prices, the AAA provided for the regulation of production for what would become sixteen commodities. The farmer who entered into an agreement with the AAA to decrease surpluses would receive benefits by adopting restricted allotments. Unfortunately, large producers responded to this arrangement more promptly than the smaller, less fortunate growers. Finally, a program had gotten through to the farmers who came to understand that problems of marketing transcended the issue of producing greater quantities of goods.

The AAA was put to the test in 1933 with a bumper cotton crop from 40 million acres. The result would be a crop of approximately 16 million cotton bales, which would be added to the surplus of past seasons. Being too late to restrict planting, the AAA identified 22,000 agents to plow up a quarter of their acreage. For doing this, the farmers would receive from $6 to $20 an acre. The agents were able to obtain agreements to take 10 million acres out of production.

Because of the reluctance of cotton farmers to sign up for restricted cotton acreage allotments the following year, Congress passed the

Bankhead Cotton Control Act (1934). A significant tax was levied on all cotton brought to the cotton gin in excess of an assigned quota. This penalty interfered with the growers' desire to take advantage of rising prices. A similar piece of legislation, the Kerr-Smith Tobacco Control Act, established taxes on tobacco. Plowing under had also been practiced for tobacco in 1933. The likelihood of a small wheat crop in 1933, gave wheat farmers a reprieve from destroying their crop.

In corn-and hog-producing states, a planned corn-crop reduction in the spring of 1934 and the devastating prospect of two-dollar hogs were motivation to kill 6 million breeding sows and hogs. This possibility revealed the seriousness of the farmers' woes in the Depression era. New Dealers thought the killing of the hogs more defensible than industry putting millions of idle workers on the roles of the unemployed. Dry salt pork (100 million pounds) was processed at government expense and provided by the Federal Surplus Relief Corporation (FSRC) to unemployed families. Federal stocks of cotton that had not been plowed under were provided to the jobless in the form of garments and mattresses. The FSRC had spent almost $300 million for such purposes by the end of 1935.

In May, 1939, a new program, termed the Food Stamp Plan (FSP), was first implemented. By the end of 1940, over one hundred cities participated in it. The FSP was designed to distribute surplus butter, eggs, fruit, pork, and vegetables. The individual on relief would purchase orange stamps. For every dollar worth of orange stamps, the individual on relief would obtain fifty cents of blue stamps free. The blue stamps could also be exchanged for surplus commodities.

In early 1936, the Supreme Court objected to the processing tax charged for surplus commodities. In a sample of ninety-six communities, taken by two sociologists, only one community responded calmly to the Court's action. The agrarian voters were upset and pressured Congress to pass the Soil Conservation and Domestic Allotment Act. The processing tax was dropped, and the emphasis shifted to conservation from crop reduction. Farmers now planted grasses, green manure crops, and legumes rather than soil-depleting crops such as cotton, corn, rice, tobacco, and wheat. Farmers were also expected to use scientific methods of farming such as fertilizing, plowing, and terracing.

This legislatively oblique approach to the problems of overproduction proved unsuccessful when in 1937 farmers experienced more bumper crops and price reductions. In February 1938, the Roosevelt administration found a way to skirt the issue. A new agricultural adjustment act was passed to address the old issue of how to regulate production continuing conservation payments to those who preferred that method. National acreage allotments for corn, cotton, rice, tobacco, and wheat were set at levels adequate for domestic use, export, and reserves. Those planting over their allotments were ineligible for parity payments or for receiving commodity loans at the favored rates. Parity was defined as price levels in the 1910-1914 period. The new agricultural adjustment act provided that those going over marketing quotas would be penalized if the quotas had been approved by a two-thirds vote of the farmers involved. Thus, the surpluses of good years were not to be dumped on the market but held to even out years of underproduction. Six million farmers participated in the program by 1940.

The effects of the AAA on the U. S. economy and the average farmer are difficult to judge. The national farm income in 1939 had doubled from that of 1932. In 1939, the national farm income was reported as $8.5 billion. This prosperity fell unevenly on the farming community. Poultry farmers and truck farmers were not covered by these programs. Particularly in the South, AAA funds had a tendency to find their way into the pockets of independent farmers and landlords. The AAA had no safeguards to protect tenant farmers against the loss of their small acreage, and good and poor crop seasons. In 1937, a tenant farmer made an average of $385, $27 of which came from AAA payments. This compared to an average cash income of $8,328 for a plantation operator, $833 of which came from AAA payments.

Now, with the element of government payments in the mix of a farmer's planning, mechanization became more possible. From 1930 to 1940 the number of automobiles owned by farmers showed almost no gain. Tractors increased by approximately 70% and trucks increased by 16%. Soon, almost 2 million of the tractors helped alter production methods and bear the heavy work of U. S. agriculture. Mechanical pickers gradually took over cotton harvesting in the South. Land was turned from cotton production to grassland to provide grazing for beef and dairy cows. New electricity assisted in this process. Meanwhile, more tenant farmers were moved off their land. In the Midwest, mechanical corn pickers came into use between 1928 and 1933. By 1939, mechanization of corn picking was estimated to have displaced one-third to one-half of the labor which Iowa's corn crop had employed. In wheat growing areas, the combine and harvester introduced in 1935 made seasonal labor in the wheat fields almost obsolete. For every 100 farm-labor jobs available, there existed 236 unemployed agricultural workers.

Farm modernization was greatly assisted by the Rural Electrification Administration (REA) which was established in May, 1935. REA offered WPA labor along with low-interest loans to cooperatives, municipalities, and states. Light and power had been denied to about one in ten farms by the refusal of commercial electric distributors to extend lines to farms. The electricity was used for multiple purposes. Heating incubators, milking, milk separation, mixing feed, pumping water, and numerous other uses were supported by rural electrification. The accelerated pace at which rural electrification was extended by REA encouraged central power plants to extend their services from 225,000 farms and farm homes in 1924 to 1.7 million farms and farm homes in 1940.

This review of the situation of farms in the 1930s reveals a paradox. The efforts to increase efficiency on farms led to overproduction and unemployment. On farms that were poor and backward, inefficiency was subsidized. In regions with larger holdings and instincts of efficiency, only the best land was used. In the Deep South, many small cotton farmers were kept at a subsistence level due to the good graces of the AAA. Ironically, acreage reduction by the AAA in corn-growing areas reduced production by only 8% percent. The adoption of hybrid corn accounted for 24 million acres in 1929. In some areas of the country, single crop farming took over. Agriculture had been adopting new technology and new farm management techniques. Fleets of tractors and gang plows were a sign of the coming of big business to farming.

The South

Although the South received only 9% of the national income, it had 21% of the nation's population. For generations, wasteful methods of farming and deforestation had eroded the South's natural assets, including its soil and woodlands. The South averaged the smallest acreage per farm of any area in the United States. At the beginning of the Depression, the South's mineral wealth and hydroelectric power remain untouched. Rural poverty was widespread, although even in urban industry the laborer received an annual average wage of $865 compared to $1,291 annually in the remainder of the United States Fifty percent of people were ill-housed. Sickness and death rates ran unusually high. More persons died without medical attention than in other areas of the United States. Overcrowded and rundown schoolhouses and an illiteracy rate of approximately 9% were characteristic of an impoverished educational system. In 1930, Georgia's average salaries for teachers were $816 compared to $1,420 for teachers' average salaries in the United States. Taxes were about half that of the rest of the country. Many of the levies fell on those least able to pay. Youth, when of age, tended to leave home. From 1900 to 1936, the South experienced a net migratory loss of 3.4 million people.

African-Americans experienced more than their rightful burden of unemployment. Approximately 2 million were on relief in 1933. The Depression severely curtailed employment opportunities for African-Americans. Jobs requiring heavy or menial labor were taken over by whites as the Depression progressed. These jobs involved unskilled industrial labor, work in the building trades, garbage collecting, ands employment as domestic help for the women. Blacks lost their jobs as whites began to compete. African-Americans faced discrimination in early stages of the relief program as well. In Mississippi, 9% of African-Americans received relief compared to about 14% of whites.

An estimated 317,000 African-Americans migrated to the North and the Midwest from 1930 through 1940. This did not compare to the 716,000 who migrated from the South in the decade of the 1920s. As they migrated north and sought jobs, the African-Americans found that there remained barriers to employment. Nearly 3 million African-Americans were supported by public funds (about one in four).

In general, the New Deal had done more for African-Americans than any other federal administration since the Civil War. Beaches (although segregated), housing projects, land-use projects providing parks and picnic areas, increased federal attention to education and health, rural resettlement, and slum clearance were among the federal largess that affected the African-American community.

The Dust Bowl

In October 1933, the Soil Erosion Service Act was passed and was established to undertake projects between the farmer and the federal government. Five hundred thirty-four of these projects were operating by 1940. The projects averaged 25,000 acres a project. A legislative act targeted at conservation was the Taylor Grazing Act of 1934. The Act authorized retiring 80 million acres of public lands from the abuse of overgrazing. Overgrazing had fostered drought and wind damage, thus restricting further homesteading on the range.

President Roosevelt issued an executive order in 1934 to create a belt of trees one hundred miles wide from Canada to the panhandle of Texas. This was designed to help conserve moisture and act as a windbreak. The program was curbed by political ridicule. Yet, many farmers were drawn to the wisdom of the program. During and after WWI, over one hundred counties in Colorado, Kansas, New Mexico, Oklahoma, and Texas had areas that were sub-marginal for farming. These areas promised profits because rainfall was better than average. Late in 1933, there was a series of storms and droughts. In the winter of 1933 and all of 1934, wind stripped the topsoil from large areas from the Dakotas to Oklahoma. The blowing soil blackened the sky at midday, buried fences and machinery, and desolated thousands of families. It was not uncommon to see bewildered families pulling handcarts piled high with whatever household goods they could rescue. In some cases, baby carriages were used to carry possessions with children walking behind. An estimated 1 million persons were transient during the worst years of the Depression. They may have been fleeing from the Dust Bowl, tenant farmers no longer needed by agriculture, or laborers now unemployed by industry. FERA's transient relief program experienced its highest monthly registration of 341,428 individuals in April 1935.

This human migration was the likely reason that the number of white farmers in Oklahoma and Texas declined. The population in the two states is estimated to have declined 6% to 7%. The African-American population of those states declined 27% and 13% respectively. The Oklahoma homesteaders (Okies) of the 1890s had been homesteading pioneers. From approximately 1935 on, these Okies and the sharecropping Arkies (from Arkansas) migrated over highway 66 to southern California where they labored in orchards, in truck gardens, and in the vineyards of the area. Over the four years beginning in midsummer 1935, 350,000 Dust Bowl farmers migrated to southern California. Many of the immigrants found unstable seasonal employment at starvation pay. Fewer than 3,000 large-scale, corporate farm operators (members of an organization termed the "Associated Farmers") employed most of 200,000 migrants in the state at that time. These farms paid below subsistence wages. The CIO attempted the unionization of labor on those corporate farms. Tension increased between the Associated Farmers and the CIO. The advance of the war industry moderated the situation due to the availability of new jobs at higher wages.

While this was transpiring, the Dust Bowl began to contract. There were several reasons that this was occurring. The belt of trees, the development of small irrigation works, building of reservoirs, the increase of farm acreage, restoration of ranching in some areas, resettlement and rehabilitation under the FSA, and federal-state cooperation were some of the measures contributing to the decline in the size of Dust Bowl areas. More than 6 million acres were severely eroded in 1935-1936. By 1939, the eroded area declined to less than 1 million acres.

Special Conditions and Programs

FERA hired unemployed nurses to look after the children of poor families. A free school-lunch program was instituted. More lasting, the Social Security Act of 1935 provided $24.75 million annually for public assistance to single parent families and their dependent children. Moreover, $3.8 million was devoted to maternal and child health. A program for homeless and neglected children received $1.5 million annually. A program for crippled children received $2.85 million. These programs, requiring state cooperation, received a strong endorsement. In 1939, Congress increased the appropriations. Between 1934 and 1938, maternal mortality rates for the nation decreased fell 25%.

In 1930, 40% of males between the ages of 14 and 19 worked. Similarly, 23% of girls aged 14 to 19 worked in 1930. Child labor had been declining gradually. By 1939, requirements for more schooling and legal restraints on child labor reduced these percentages to 35% and 19% respectively.

In 1934, agriculture continued to employ children between ages ten and fifteen. Children of southern sharecroppers worked for landlords from the age of six or seven forward. Child labor increased slightly after the invalidation of NRA in 1935. The Walsh-Healy Act in 1936 barred those holding sizeable government contracts from hiring individuals below sixteen years of age. In 1937, sugar-beet farmers receiving government benefits were also prohibited from using underage labor. The Fair Labor Standards Act, in 1938, banned industrial work by youth under sixteen. The Act forbade employment of those under eighteen in hazardous or injury-prone occupations. The Act offered no protection to the non-factory worker and the laborer in intrastate businesses. Child labor laws were favored partially because there was an adult labor surplus. A child labor amendment was proposed as an amendment to the Constitution. As might be expected, management interests in states employing child labor lobbied against this reform. Ultimately, the amendment saw its demise at the hands of the Supreme Court in June 1939.

Youth experienced more unemployment than any other part of the labor force. In 1935, the American Youth Commission estimated that there were 4.2 million unemployed youth. This represented over one-third of the nation's idle manpower. Part of the unemployed youth took to the road. On the Missouri Pacific, the number of bums and freight-car migrants rose from 13,000 in 1929 to almost 200,000 in 1931. Free flophouses and midnight missions in Los Angeles gave refuge to more than 200,000 individuals. In the summer of 1932, railroads, rather than trying to inhibit train-hopping, put one or more open box cars on their trains to prevent Individuals from breaking into sealed cars. Policemen, railroad employees, and social workers were in agreement that the great majority of these transients were not criminals or vagrants.

Crime reached its high point in the period between 1932 and 1934. Former rum-runners and others involved in activities of waning profit turned to kidnapping children and adults. Congress passed laws with severe penalties. These laws prohibited interstate abductions (1932 and 1934) and provided severe penalties if the victim was harmed. For cause of death rates in three categories, see Table 3.3. Homicide rates increased from 8.4 per 1,000 in 1928 to 9.7 per 1,000 in 1933, suicide rates increased from 13.5 per 1,000 in 1928 and peaked at 17.4 per 1,000 in 1932. Unknown causes of death rose from 13.5 per 1,000 in 1928 to 14.4 per 1,000 in 1930. Homicides per 1,000 was the only rate that tapered steadily after 1933 to 1940. The years of 1937 and 1938 represented increases in a general pattern of decline in the suicide rate. The decline in unknown causes of death continued after its high in 1930.

Table 3.3: Homicide Rates, Suicide Rates, and Deaths from Unknown Causes/1000 1928-1940*

Year	Homicides/1000	Suicides/1000	Unknown Causes per thousand
1928	8.4	13.5	13.5
1929	8.6	13.9	14.0
1930	8.8	15.6	14.4
1931	9.2	16.8	12.8
1932	9.0	17.4	11.9
1933	9.7	15.9	11.5
1934	9.5	14.9	10.9
1935	8.3	14.3	10.5
1936	8.0	14.3	11.1
1937	7.6	15.0	10.2
1938	6.8	15.3	9.8
1939	6.4	14.1	9.3
1940	6.2	14.3	9.9

* Vital Statistics Rates in the United States, 1947, Linder and Grove

Other Roosevelt Programs

The CCC was one of the most expensive programs per participant. The program cost about $1,075 dollars per year to maintain each participant. Offsetting this cost, cabins were constructed, fish and game sanctuaries were designated, lakes were made, parks were enhanced, roads and trails were built, forest fires were checked, erosion was stopped, two billion trees were planted, and wildlife was preserved. Additional gains in the health and self-respect of the participants were made.

Nine states created new parks with federal assistance. Between 1933 and 1936, 600,000 thousand acres were added to state parks. Purchase of lands for parks and wildlife refuges increased from an annual pre-Roosevelt average of about 500,000 acres to 2 million acres in 1935. The public approved. Travel statistics revealed that 6 million visitors came to parks in 1934. This increased to over 16 million persons in 1938.

In June 1935, the National Youth Administration (NYA), designed for young men and women between ages of sixteen to twenty-five years) came into being. It acted as a junior WPA. It was part-time work relief for this group. Its cost was the cheapest form of work relief at an annual amount of $225 dollars a participant. In its peak month (April 1937) NYA had 600,000 in its two major programs. Seven-eighths received student aid, and the rest were employed in out-of-school projects. The latter group lettered street signs, repaired discarded toys for children, built school furniture, constructed footbridges for rural paths, sewed, and participated in soil erosion control projects. The larger group of beneficiaries (approximately 2 million) consisted of high school and college students who required financial assistance to help in continuing their education. Enrollments in colleges experienced a steep drop in 1932 to 1934. In February, 1934, FERA initiated a program of financial aid to address this issue. FERA supplied approximately $15 million a month for 75,000 young males and young females. The NYA inherited the program and greatly expanded it. After the NYA took over, the young people were involved in tasks such as assisting in campus maintenance and repair, cataloguing and mending library books, and compiling statistics. The NYA projects depended heavily on local management. This often brought problems.

Public schools experienced serious problems throughout the Depression. Schools were often the victim of municipal economy drives. In Chicago, teachers were unpaid with the issuance of scrip and tax warrants prolonging the agony. In New York City, 11,000 teachers did not work in 1932 and 1933. At the same time, five out of six Alabama schools did not open. By the outset of 1934, 2,600 schools (primarily rural) had closed their doors. These conditions gradually improved with the upturn in the U.S. economy.

The spread of adult education is worthy of special note. In the fall of 1933, the Federal Emergency Adult Education Program was established. Using unemployed teachers, the program instructed other groups of the unemployed. In April, 1935, there were 43,722 teachers and 1,190,131 students. The majority of students were literate, able-bodied adults. There was federal encouragement for the establishment of special colleges and junior colleges to attract additional unemployed persons. Despite these efforts, numerous architects, chemists, young doctors, engineers, and lawyers graduated by their alma maters, found themselves unemployed. By 1933, the unemployed graduates included some 200,000 certified teachers. They had to neglect their training and professional pride to take jobs that became available. Often, this included a spell on relief or labor in fields or orchards. It was not until 1940 that registration in forty-four state universities and land-grant colleges totaled about 1.5 million, a national record.

The Consumer

In 1934 to 1935, the National Resources Committee, assisted by the WPA, conducted a survey. They surveyed consumer units or individuals living alone. With an average income of $1,500 dollars, two-fifths earned less than $1,000, and one-third less than $780 dollars. Because the lowest third spent $1.207 billion more than they earned, one in three required some type of relief. At the high end of the economic scale, one consumer unit in eight had an income of about $2,500. One in thirty consumer units had an income of $5,000 or better, and one in a hundred had an income of $10,000 dollars or higher.

The nation had approximately $50 billion at that time for consumption. Citizens spent approximately 34% on food, 19% on housing, 10.5% on clothing, 10.67% on household operation, and 2% for household furnishings and equipment. In addition to these basic needs (76.17%), personal care required $1 billion, automobiles about $3.8 billion, other forms of transportation about $884 million, recreation $1.6 billion, and tobacco $966 million, and reading material required $551 million. Private expenditure for education was $506 million. This was greatly supplemented by government and endowed institutions, raising the former outlay to a total of $2.4 billion.

Medical care was a cause for concern. Medical care cost 30 dollars per year per capita. Individuals having incomes between $1,200 and $2,000 per year spent an average of $13 per year. Persons with incomes of less than $1,000 per year paid $9 for medical care per year. These two groups represented 50% of the U. S. population. In 1932, medical care cost $30 per capita per year. Twenty-three dollars per capita per year came from private pockets. The remainder came from government and philanthropy. The penalty for a year of bad health could make the difference between solvency and a long period of debt and worry.

In 1932 to 1933, investigations revealed that the highest sickness rates were among wage-earning families who had experienced abrupt losses in income and declines in living standards. The incidence of illness was approximately 40% higher among the jobless than among the full-time employed. Children and youth remained the primary victims. From 1940 to 1941, Army medical examiners processed approximately 2 million recruits. About one-half of these individuals were rejected. This resulted in the Army lowering its standards.

In 1937, the Resettlement Administration pioneered federal efforts to foster cooperative self-help with localities. It assisted in setting up dental, hospital, medical, nursing, and surgical services among drought-stricken and poor farm families in the rural areas of the Dakotas. By January 1940, over one-third of a million individuals were covered by county health associations and other cooperative arrangements developed in thirty states by the FSA and local physicians. The National Cancer Institute was formed in 1937 for cancer research. In 1938, a Venereal Disease Control Act passed. It provided for the inspection, prevention, and treatment of venereal disease. In 1940, federal public health activities were placed in a new center for the National Institutes for Health at Bethesda, Maryland. The establishment of private insurance to cover medical costs brought families with moderate incomes health insurance coverage. This plan provided reimbursement of health costs to individuals paying a set monthly rate for hospital care or complete health coverage.

Aging, Pensions, and Social Security

By 1934, FERA provided temporary assistance to approximately 750,000 persons sixty-five and over. A minimum of 1 million were receiving public relief by 1936. Three-quarters of a million remaining received assistance from their children, friends, or relatives. The clamor for state pension systems became so strong that by the middle of 1934, twenty-eight states had passed legislation for the aged. Twenty-three of these were mandatory.

In 1934, President Roosevelt named a committee to prepare a program that became passed as the Social Security Act on August 14, 1935. The Act provided two types of assistance for the aged. The Act provided for what was, in effect, an annuity of sorts. Employees and employer-matched funds for contribution to the Social Security Trust Fund were required. Participation was compulsory, with the exception of federal government employees, casual laborers, domestics, merchant seamen, and employees in charitable, religious, and educational institutions. As of 1940, 52 million persons had acquired Social Security numbers under the program. Citizens could retain income from investments and savings. Employment above a certain amount resulted in federal taxes being increased.

The second type of assistance provided grants for persons who had aged past their years of income production. The federal government cooperated with states for subsidies for the indigent old-aged. The amount to be provided could be up to $30 a month (later changed to $40 per month). Payments varied from this amount with some Southern states offering small sums, and California providing close to the maximum. Participation was widespread. In 1940, 2 million elderly poor were enrolled in state systems. In addition, there were 50,000 blind and 900,000 disabled, neglected, and dependent children receiving similar benefits.

An important part of the Act was its provision for unemployment insurance. To finance the required reserve for this program, federal payroll tax of 3% was levied on employers having eight or more employees. The previous exceptions to Social Security applied. The states administered the system in large part, while the federal government created the major rules and funded administrative costs. Benefit payments varied based on previous earnings and length of employment. Initially, benefits were paid for a fourteen-to-sixteen week period. Benefits were not available to those who voluntarily left work, were fired for misconduct, refused to take a suitable job, and, in some states, had been striking.

The Tennessee Valley Authority

One of the crowning achievements of the Roosevelt administration and Congress was the Tennessee Valley Authority (TVA). In 1918, Woodrow Wilson had constructed a dam and two nitrate plants at Muscle Shoals on the Tennessee River. The two plants were to make explosives to be used in war and peace. The area drained by the Tennessee River included seven states with a population of 4.5 million. Roosevelt, wanting to see the power resources developed for the people of that area, created the TVA for electric power production, flood control promotion, proper land and forest use, and the economic and social well-being of the people. A planning council had six divisions: agriculture, engineering and geology, industry, land use, social welfare, and economic welfare. While building locks, dams, and power plants, the TVA started working on reforestation, retirement of sub-marginal lands, soil conservation, promoting the use of better farm machinery, the promotion of local business, education, and public health.

The region saw a radical transformation that soon became the pride of its citizens. The TVA provided 9,000 miles of shoreline for recreation, with the waters stocked with fish. The Tennessee Valley Waterway Conference cooperated with TVA experts to prepare plans for a group of public-use terminals connecting railroads and truck highways. A navigable channel of 600 miles was created. The proof of the value of the TVA construction came in January 1937, when an Ohio River flood drowned about 900 victims and left approximately 500,000 homeless.

Torrential rains were weathered by the TVA and the Tennessee River without incident. In 1940, the ammonium nitrate plant at Muscle Shoals went into full-scale munitions production, thus, achieving President Wilson's initial objective.

Electrical power was supplied to consumers at the rate of three cents per kilowatt hour. This compared to a price of ten cents a kilowatt hour in other areas of the United States. The consumer was now able to purchase electric pumps, hay driers, freezers, motors to grind feed and cut wood, and more. The availability of power stimulated the production and consumption of household goods also. Despite all of these benefits, the producers of private power disliked the new agency. Private electrical rates were established to remain competitive with the TVA rates. Consequently, some of the facilities of the Tennessee Electric Power Company were sold to the TVA and other utilities.

The U.S. Congress proceeded to create six more projects targeting the Atlantic Seaboard, the Great Lakes, the Ohio Valley, the Missouri and Red rivers, the Columbia River basin, the Colorado River and Pacific Coast, and the Arkansas Valley to the Rio Grande. President Roosevelt delivered enthusiastic speeches in early October 1937 on the sites of the Grand Coulee dam and at Fort Peck, Montana, lauding these projects. Two days later, Roosevelt's speech in Chicago addressed aggressor nations.

The 1938-1939 Recession

Because of the 1938-1939 recession, President Roosevelt and Congress shifted their spending program back to programs that would stimulate the economy. They increased appropriations by $5 billion to support the programs of the WPA, PWA, and the lending activities of the RFC. The potential misery for some people was blunted by Social Security.

Modifications of the Social Security Act occurred. Benefits were added for survivors and dependents of Social Security beneficiaries. The United States Employment Service joined the Social Security Board. Job insurance and job placement were combined. The number covered by unemployment insurance increased until 1940. Soon after that, the availability of defense jobs and better wages caused people to be less willing to retire at sixty-five. The Federal Security Agency became the primary instrument of federal social welfare activities. It included the CCC, the NYA, the Public Health Service, and the Office of Education.

Even the opposition had come to recognize the value of Social Security. A 1938 Gallup poll found that nine out of ten respondents favored old-age pensions. Social Security was here to stay.

The Coming War

As the dangers of war approached, the WPA focused increasingly on projects such as aircraft housing, armories, camps for the National Guard, housing for munitions workers, and rifle ranges. By October 1941, one-third of WPA workers were involved in such tasks. By early autumn 1940, the Federal Reserve Board's index of industrial production had increased by seven points. Approximately 2 million more persons had found private employment. The first twelve months of war had passed in Europe. A different demand for labor came into being when the first peacetime Selective Service Act was signed on September 16, 1940. Along with this came the drama of another presidential election year.

The Republican nominee, Wendell Willkie, had differences with Roosevelt's methods, not his aims. The Democrats drafted Roosevelt for a third term. Willkie's stands were much more clear-cut than those of his party. In 1940, Willkie strongly endorsed selective service. Eight Republican senators had voted yea, to fifteen nay. Fifty-two representatives for his party voted yea and 112 voted against selective service. After the election had come and gone, Chief of Staff George Marshall made a personal appeal for support of the Selective Service Act. Despite this, the House of Representatives passed the legislation by one vote (203 to 202). The conservative old guard of the isolationist Republican old-guard was not at ease with Willkie, because of what some interpreted as his unpredictability and his resistance to bossism.

A Gallup poll in 1937 asked whether the United States should join a world organization with police power to maintain world peace. One out of four responded in the affirmative. By the autumn, two out of five agreed. The next year witnessed the number grew to three out of five. With an embargo on the shipment of material to Japan and the freezing of Japanese assets in the United States, the opposition to the Axis powers began.

Summary

While all New Deal programs were not a success, the PWA and WPA had some effect on economic conditions. Throughout the era, there remained between 4.3 and 12.8 million unemployed persons. Agriculture received special attention receiving extensive federal subsidies. The extensive reforms of Roosevelt's administrations drew much popular attention and attracted the following of Members of Congress and local politicians. Others were startled and often deeply concerned to learn that one-third of the nation was poorly housed. A more responsive social conscience was cultivated by the attention called to sharecroppers, slum dwellers, sub-marginal farmers, sweat-shop labor, and underprivileged families. In response came a variety of reforms. The era of the Depression had started under the leadership of President Hoover. His belief in voluntarism and individual initiative, plus the support of corporate and business interests proved to be a weak antidote to the serious, extensive, and complex problems being confronted by the United States. [1]

---

[1] Data in Chapter III are taken from the Historical Statistics of the United States; Wecter, 1948, Linder and Grove, 1947; and Hosen, 1992.

CHAPTER IV

A CONCEPTUAL OVERVIEW OF THE GREAT DEPRESSION ERA

The stages of monetary and fiscal crisis are characterized by Charles Kindelberger in his book, *Manias, Panics and Crashes: A History of Financial Crises* (1989).

The first stage he identifies as a financial boom. This was definitely true preceding the Great Depression. The boom was marked by a rapid expansion of credit. The expansion of credit came prior to the market Crash in September and October 1929. Expanding personal credit through the existing system of banks, creation of new banks, establishment of new credit instruments, and development of personal credit outside the system of banks were features of the boom and the Crash. This mixture of credit sources provided an important source of financial instability by making methods available to finance the boom.

The second stage presented an opportunity for speculation that was increased by availability of credit. Demand for goods or financial assets increased. During the boom from 1928-1929, the rapid increase in share prices offered increased opportunities for profit-taking. Increased numbers of individuals, investors, and firms participated in the stock market. When increased investment leads to increases in income, euphoria develops. This euphoria results in speculation to buy into price increases in investments.

The third stage is identified by individuals getting caught in the speculative fever by buying on installment or on margin. Increasing number of individuals and firms get involved. At some time, investing becomes irrational in anticipation of increased rates of return.

In the fourth stage, the credit system becomes extended to its limits and beyond. Prices, interest rates, and the velocity of money increase. The time comes when a number of insiders catch on to what is occurring and sell out their holdings.

Lastly, the panic begins and eventually works its way through the financial markets and the economy. Those who have purchased on margin cannot cover their loan obligations by selling their holdings. If they do sell, the proceeds from assets of declining price do not meet obligations, and bankruptcies or the elimination of assets, wealth, and fortunes occur.

In the U.S. stock market, share volume declined from 11.25 billion shares in 1929 to 208 million shares in 1940. This was a decline in trading volume of 81.5% over that time. The decline was 11.042 billion shares, or an average decline of over 1 billion shares a year. The magnitude and direction of the contraction sent shock waves through the U.S. economy. Whether these shock waves were responsible for the Great Depression era provides some of the subject matter for the rest of this book (see Figure 4.1).

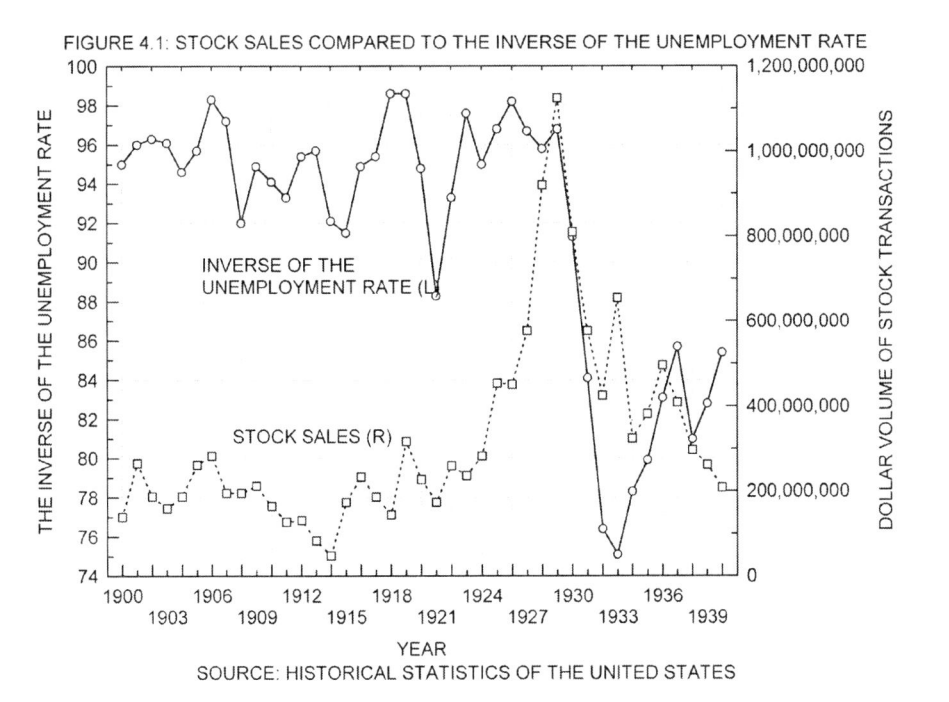

FIGURE 4.1: STOCK SALES COMPARED TO THE INVERSE OF THE UNEMPLOYMENT RATE

SOURCE: HISTORICAL STATISTICS OF THE UNITED STATES

There was a general downward trend in the number of shares of stock traded from 1929 to 1940. A steep decline occurred from 1929 through 1932. This was accompanied by an increase in 1933 and a lesser increase in shares traded from 1934 through 1936. The general decline from 1929 through 1940 with its irregularities provides little evidence of a relationship with the post-1933 tendency towards recovery. The decline from 1929 through 1932 may, however, have had something to do with the lack of confidence that developed over that time period in the financial institutions of the United States. Table 4.1 provides the classification of downturns in the U.S. as completed by the National Bureau of Economic Research's Dating Committee.

Table 4.1: Business Cycle Dates and Periods of Duration*

Business Cycle Reference Dates		Duration in Months	
Peak	Trough	Cycle	
(Quarters in Parentheses)		Trough from Previous Trough	Peak from Previous Peak
June 1899 (QIII)	December 1900 (QIV)	42	42
September 1903 (QIV)	August 1904 (QIII)	44	39
May 1907 (QII)	June 1908 (QII)	46	56
January 1910 (QI)	January 1912 (QIV)	43	32
January 1913 (QI)	December 1914 (QIV)	35	36
August 1918 (QIII)	March 1919 (QI)	51	67
	July 1921 (QIII)	28	17
	July 1924 (QIII)	36	40
October 1926 (QIII)	November 1927 (QIV)	40	44
August 1929 (QIII)	March 1933 (QIII)	64	34
May 1937 (QII)	June 1938 (QII)	63	93

* Taken from *http://www.nber.org/cycles.html/*

CHAPTER V

SCHUMPETER AND CAPITALIST EVOLUTION

Introduction

As an example of external factors affecting business, there are political and natural events. These events are used to diagnose a business situation or predict the future. Joseph Schumpeter (1939) maintains that most every businessman realizes that his business is dependent on conditions over which he has no control. These conditions make up what is termed the general business situation. There is something else that affects the fortunes of all businesses and is more than the total of the merits of individual firms. It is Schumpeter's purpose to interpret this general pattern of economic life. The question is posed: do the changes provide any significant regularity? A normal year has been had by businesses that earn enough to pay current expenses, depreciation, and contractual interest on debt, although not enough to increase or decrease their investment more than normal. Most businesspersons use this type of comparison to gauge whether their particular business has had a "normal" year.

It is important to note that businesspersons actually compare their current situation with the normal. The businesspersons rely on various information sources available to them. There may be some rules or facts upon which the businesspersons base their judgments. The businesspersons try to diagnose economic change and its causes.

Among those that act on an economy from outside are natural disasters such as the great tsunami in southeastern Asia. These phenomena are usually unable to be accounted for within the sphere of economics.

For Schumpeter, it is important to distinguish events resulting from the workings of the economic system from those acting on the economic system.

Schumpeter maintains inventions and discoveries create new possibilities and are among the most important causes of economic and social change. The discovery of the United States achieved relevance only when new possibilities were converted into commercial and industrial reality. As inventions were put into practice, a process that is part of the economic life of the inventions takes place.

Schumpeter's vision of the process of capitalist evolution included the opening of new markets, occurrence of inventions, and mechanization of industry. He observes that technological progress accounted for the rate of increase of production in the nineteenth century. The very essence of capitalism is its promotion of technological progress. The two cannot be separated.

Migrations are conditioned by business fluctuations. Variations in numbers and age distributions due to causes other than migration, as well as migration, will for purposes of his writings be disregarded. It is difficult to demonstrate relations between the rate of change of population, nativity, and mortality on one hand and economic fluctuations on the other.

There are also changes in the institutional framework such as tariff policy and taxation. They include such items as the programs of the New Deal and noticeable changes in social behavior and habits. An example of the former includes changes in the responsibilities of a Central Bank. An example of the latter includes changing from keeping funds in a bank in the form of demand deposits to hoarding and keeping resources in the form of cash. Such changes do not have to occur in the form of legislation. Their importance is that they change the rules of the economic game, the significance of economic data, and the systematic relationships that constitute the economic world.

Our economic system is not pure. It is always in transition towards something different. Because of this, logical models of the economy cannot always be used to describe our economy.

In fact, external factors are numerous, and may account for a great deal of the variation in an economy. One can wonder if these external factors account for the wavelike alternating states of prosperity and depression that are experienced. Schumpeter maintains that some of these disturbances occur at nearly regular intervals to introduce a process of adaptation in the economic system that generates a wavelike variation in every case. There are times, covering considerable amounts of material in Schumpeter's *Business Cycles* (1939), in which external factors overshadow and overpower everything else.

It is unreasonable to think that the outline of the phenomenon presented can be derived from statistical material alone. All that can be said is that the contour lines are irregular. What we know is not the working of capitalism, but of a distorted capitalism that reflects variation both past and present. For example, the fundamentals of business and industry have been politically shaped.

In some cases, it may be possible to gather enough information about the duration, nature, and extent of disturbances to understand which figures are affected by it. To understand time series it is very important for U.S. to thoroughly understand the economic history of the time. This knowledge may extend to the level of a particular industry or firm. Cultural, economic, political history, social history, and industrial history are essential to our understanding of the issue of business cycles.

The businesspersons' or industrialists' best strategy is to survey a broad range of factors, both internal and external, in making business decisions in particular, indices that are sensitive to a particular business or industry. This search of the researcher is one in which he/she needs to seek out symptoms of causal significance rather than those that characterize the symptoms of factors resulting from business or industrial action. In practical terms, rational opinions are sometimes inadequate to explain the fluctuation of industrial factors. The primary role of semiology (the study of signs and signals) is to diagnose the economic state of a country, not just economic fluctuations.

The facts and data that must be judged in their whole include: consumers, or household demand for goods and services; expected profits and actual profits; the demand of producers for goods and services; wholesale and retail commodity prices; bond rates and money rates; employment and unemployment; bank clearings and debits; bank and business failures; the quantities and values of exports and imports; total production including finished goods, finished equipment; finished consumer goods; semi-finished metal products; iron and steel; electrical power; unfilled orders; bank and business failures; securities issued; stock prices and shares traded; consumption and shifts in categories of consumption; the total of money income; differences in income strata; stocks of finished products or commodities being accumulated by producers, merchants, or other manufacturers; sales by chain stores, department stores, mail-order hoU.S.es; the reserve ratios **of** banks; bank loans; demand deposits; consumer and business loans; broker loans; car sales; wage rates; new firms; liquidations; actions and attitudes of creditors, customers, and the banking community; gold flows and production; factors indicative of railroad activity; the nature of the real estate market; particular categories of receipts and expenditures of federal, state, and local governments; foreign exchange rates; proportion of industrial capacity active; dividends; the mood of the business community; the marriage rate; the rate of family formation; internal migration within the United States; immigration and emigration; advertising; excess production; velocity of money; and, hoarding.

This listing is not complete, showing the complexity of economic analysis. In many instances, there is a discrepancy between what we wish to infer from a set of data and what it actually means. Thus, great care must be taken in the interpretation of data. Made necessary and permissible by the purpose of Schumpeter's work are seasonal adjustments that have not been entirely successful or do damage to the material used. Points will be argued as if this had been successfully done. Doing so is an important part of the simplifications made permissible by the purpose of this book.

When a time series is corrected for seasonal variation and for trend, its units display a wavelike form. Such waves are known as cycles. These cycles represent a distinct form of historical reality that includes disturbances or changes due to external factors. By applying multiple time series to the same chart, comparison is made more productive. Impressions can be formed by observing and measuring the periods, amplitudes, and timing of the movements of each time series relative to the others. This method proves especially useful when measuring antecedents and lags.

Some series indicate the entirety of a business situation. These will be termed *systematic* series as contrasted with *individual* ones that record conditions that are part of a business situation. Price levels and total production are examples of systematic series that are synthetic. Natural series may be systematic also. Some examples are unemployment and interest. Variables may be *causal* or *consequential*. In addition, they may be *primary* or *secondary*. The two distinctions may cross as a primary indicator may be consequential, and a causal indicator may be consequential.

Very reasonable judgments can be made by being careful about one's facts and conclusions. Furthermore, the interdependence of many of these factors is evident. Numerous factors are related to national income, and in turn, are shaped by national income.

It is sometimes very difficult to determine the direction of causation. It leads to some bewildering questions. For example, does national income lead to consumption, or is it consumption that leads to national income?

Much can be done with the facts and others that are indicative of the business situation. Schumpeter's book, *Business Cycles*, uses a range of facts and a defined methodology to analyze historical economic situations as well as the situation up to 1939. In his analysis, he indicates that there is no such thing as a single cause or set of causes that accounts for booms, crises, and depressions.

Equilibrium

Economic life need not be treated as a static set of circumstances. It is best viewed as a process of organic growth adapting itself to changing circumstances and data. Schumpeter views the economy as a number of waves that are a result of this adaptive process. The waves themselves do not point to an internal cause of business cycles. Some have held the cycle to be spurious or random. The task that remains is to construct a model of the economic process and see whether it fits the operation of time series. It therefore follows that we cannot assume that there is a cyclical movement as an inherent part of the economic process.

Schumpeter states prices and quantities are interdependent and form a system. The most important task of economic analysis is to examine the properties of that system. What we desire to learn is whether or not the known relations and the data are sufficient to determine the components of price and quantity. The values of quantities and prices that are the ones that satisfy the interdependent relationship are said to be *equilibrium values*. A system that exists if all of its prices and quantities are in equilibrium is said to be in a *state of equilibrium*. Also used are the self-explanatory terms, such as *stable, indifferent, or unstable equilibrium*.

This type of general equilibrium implies that every firm and household in the economy is, by itself, in equilibrium. For firms, this indicates that under present conditions, no firm is able to increase its revenue by shifting any component of its monetary resources (capital) from what it is spent on to any other component. Similarly, for households, given existing circumstances, no household feels it is able to improve its existing circumstances by transferring any part of its money income from how it is now actually spent to some alternative way of spending it. More generally, all firms and households believe that under existing conditions, by examining the components of their economic situation, they can determine what is within their power to change, and whether or not they can enhance their position by altering their economic behavior.

Some of the conditions that exist in these circumstances are: every firm's and every household's budget must balance; all quantities produced by firms must be purchased by other firms or households; all existing factors of production must be used to the extent that their owners would like to see them used at existing prices; and no demand goes unsatisfied at existing prices.

When general equilibrium prevails, every firm, every industry, and every household is in equilibrium. For some purposes it may be useful to talk of partial equilibrium. In such circumstances, an individual firm or industry may be in equilibrium while others are not. The concept of partial equilibrium may be accompanied by a system of related aggregates such as total output, total income, and total net profit. If these components are used so that they do not have a tendency to change coming from their relationships to one another, then *aggregative equilibrium* may be discussed. It would be misleading to discuss aggregative equilibrium as if it could create change by creating disturbances in the economic system as a whole.

Much reasoning about aggregates is faulty and makes for superficial analysis. One of these superficialities is the theory of monetary cycles. One of the relationships subsisting between aggregates is the quantity theory of money. In fact, it is only one condition of equilibrium. Other conditions are the microanalysis of the components of aggregate measures used to characterize an economy.

Another concept is introduced that is of special importance to general equilibrium. If the components of an economic system exactly satisfy the conditions and relations constituting the economic system, it is said to be in *perfect equilibrium*. The term and conditions known as *imperfect equilibrium* are clearly contrasted to the conditions of perfect equilibrium.

What is of importance is whether or not there is a disposition of the economic system to move towards equilibrium. If the concept of equilibrium is to be useful as a tool of business cycle analysis, the economic system would respond to a state of disequilibrium by attempting to restore equilibrium in the face of such change. The concept of equilibrium discussed by Schumpeter indicates that equilibrium is an approximation that is short of what is needed for understanding the processes of a constantly changing economic world.

Later, patterns of reality may be discovered that require abandonment of the general equilibrium model. Economic behavior is not satisfactorily explained in terms of the values that variables attain at a given time. For example, quantities of demand or supply are not a function of prices prevailing at a time. The impact of past and expected prices must also be included. This type of variation is termed *dynamic*.

A case in point is that of technological change and lags. There are always components in the setup of a firm as well as in an economic system that cannot be adapted too readily. There are other changes that can be quickly adapted to. Full or perfect equilibrium may take much time to happen, if it happens at all. New disturbances are always affecting an imperfectly equilibrated system. This fact does negate the presence of a disposition toward perfect equilibrium that asserts itself and assists in explaining many processes. This is all that is desired.

Lags may result from causes other than technology. Frictional unemployment and industrial change from the production of one quantity to a different quantity are examples. The introduction of friction into a situation will always mean equilibrium different from what might otherwise have been. This is further complicated by different components or different sectors of the economy experiencing different levels of friction. Lack of balance will follow as components of the economy get out of balance with each other. Friction is not an entirely negative force. Some friction may be needed for the functioning of an economy. Friction has the function of steadying the adaptation of the economy to changing circumstances. Such friction may be essential to the economy's functioning.

Situations of frictional resistance to change are sometimes termed as stickiness or rigidity. There is some difficulty in measuring these terms. The discussion of these factors is of a non-technical nature. There may be an effort to introduce stickiness or rigidity if, for example, a price is regulated by public authority, or a business, or industry. It may be the purpose of that public authority, business, or industry to stabilize the price in question. In such a case, the price may represent stickiness or rigidity relative to other prices of similarly priced products.

Growth and increasing demand, decline and lack of demand, insert themselves into the economic process. Business and industry rely on accurate expectations about growth and decline. This remains one of the most difficult tasks of management, to estimate and match demand. In other terms, it involves establishing equilibrium between supply and demand. The penalties to the businessperson or industrialist are very noticeable. The status of inventories is a tell-tale clue. Growing inventories may indicate an imbalance of supply and demand. Declining inventories may indicate an excess of demand over supply. Finding equilibrium appropriate to a business or industry is a continuing process of managing inventories, and regulating supply to satisfy demand.

Under perfect competition, there is a disposition towards equilibrium. In the world of business cycles, the study of equilibrium can be useful. In perfect competition, the individual business or industry may be powerless to change the market price, but feels a need to accept it. The business or industry risks losing business if a higher price is charged. By charging a lower price, the business or industry may be subject to a loss that will, in the long run, threaten its existence.

The monopolist faces different circumstances. If the monopolist chooses to charge higher or lower prices than a price that maximizes his or her profit, that individual would suffer consequences in that, within limits, his or her gain would be less than under a maximizing paradigm. Consideration of public opinion may lead the monopolist to abandon maximizing immediate gains for long-term gains. The monopolist may respond to the issue of how to obtain maximum gains in any of several ways. Each of these options provides a determinate result and yields an equilibrium mechanism.

One of the ways of accommodating the balance and near balance of the U.S. economy is to speak of an equilibrium zone. This range, surrounding an equilibrium point, is a more accurate characterization of an economy's balance due to the unlikelihood of achieving a perfect equilibrium. An example of an equilibrium zone occurs in bargaining between management and labor. The strategy between both sides is to vary wage rates and hours by small increments without having to bluff. There are a wage rate and number of work-hours which are to the advantage of business. Similarly, there are a wage rate and level of work-hours that are supported by labor. These wage rates and levels of work-hours will differ between management and labor. Their difference provides a zone of indeterminateness. Such an equilibrating mechanism works within the specific conditions of each case.

Circumstances change so quickly that assuming perfect knowledge and a fixed reaction is not permissible. The characteristics of changing circumstances provide some of the information needed to narrow indeterminateness. Temporary requirements, planned strategy, and changing anticipation of events require a wider scope of analysis. This yields zones and shifting zones of equilibrium. In addition, demand and supply curves may not be independent of each other.

Equilibrium theory provides Schumpeter with the basis for some of the logic necessary for his analysis. The lack of precise definitions for overproduction, excess capacity, and maladjustment makes the concept of equilibrium relationships much more necessary. The construct of equilibrium, although such a state may never be achieved, is useful as a point of reference, and for analysis. Actual conditions can be measured in relation to the equilibrium state. It is sometimes a practice of comparing actual with normal values. This approach is not invalidated by the use of the equilibrium concept.

Evolution and the Economic System

There are both internal and external factors of change that affect the economic system. Included in the category of internal change factors are considerations such as changes in quality or quantity of production factors, changes in tastes, and changes in means of supplying commodities.

Schumpeter maintains that concepts of consumers' utility functions or indifference varieties are negligible in the changes they introduce into the economic system. Most changes in the tastes of consumers are brought about by producers' actions. For example, railroads have not developed because consumers initiated effective demand for the service. There are other examples. Consumers did not initiate the development of travel by the automobile or airplane. Similarly, production of radios, electric lights, chewing gum, or rayon stockings were not initiated by consumers. Consumers did not conceive of their existence.

There are some exceptions to this. In certain instances, there are fashion leaders who create new additions and changes of habit to private lives. There are also changes or movements that affect the consumers' goods bought by households. Change in temperance laws is an example. Changes in demand that come about in that way usually involve different choices among existing commodities. If such changes are unsupported by changes in disposable personal income, business and industry usually adapt themselves. Whenever exceptions, such as war, occur, they can be treated by considering their special conditions. Such circumstances are not included in Schumpeter's view.

Population and Economic Evolution

Changing productive resources may appear to be the prime mover in the process of internal economic change. Holding the physical environment as a constant (with the development of new nations as exception), the increase in productive resources may well be the result of an increase in population and an expansion in the volume of producers' goods and services. The listing of population change as an external factor is that no relation can be established between variation in population and variation in the flow of commodities. A current (1938-1939) demonstration of this was found with the populations of China and India. Increase in population does not bring about change other than a reduction in disposable personal income per head.

Concepts to Be Used

Savings means the designation, by a household, of a component of current income for use in acquiring titles to income or for debt repayment. A business or industry doing the same thing with a component of its net receipts from the sale of goods or services is termed "*accumulation*". The difference between *savings* and *accumulation* also applies to other situations where it may be difficult to make the distinction. In cases such as farming, household and "firm" may be the same. Both saving and accumulation will be taken as components of a monetary process and will not be defined in regards to goods and services. The concept of *dissaving* (including the spending of capital gains) and *decumulation* are opposite in meaning to saving and accumulation respectively.

The gathering of a sum of money designated for a purpose such as buying a consumer durable good or covering an expenditure that cannot be met out of current receipts is not included in Schumpeter's meaning of saving. Saving does not include the rearrangement of the time shape of one's disposable personal income. Thus, the choice to save is not to spend or defer spending, as these actions likely mean later consumptive use of the funds.

Investment shall be defined as deciding and carrying out the process of acquiring titles to income. For household, the purchase of bonds, stocks, mortgages, and land and buildings intended for business use all constitute acts that shall be termed *real investment*. In the instance of firms, real investment includes spending on all types of producers' goods while putting aside the issue of replacement. Savings and investment as defined here are separate events. Both remain as internal factors that can affect economic change.

The term *growth* shall be used to mean changes in population (also in age distribution) and in the total of savings and accumulations corrected for variation in the purchasing power of the dollar. The term is used to emphasize that the variation in these two variables is continuous. Schumpeter maintains, as an example, that population is a function of time such that the function has a finite value equal to a limit as the variable time approaches a chosen point. Population change happens at a slow rate that is incapable, by itself, producing the dramatic changes in business and industry that are witnessed. This does not mean that variation in population cannot cause some fluctuations, it surely can.

Schumpeter maintains that innovation plays an important part in economic life. He defines it as changes in the methods of supplying goods and services using a much broader definition of the phrase. The introduction of new goods and services, technological change, opening of new markets, developing new sources of supply, improved handling of materials, applying Taylor's methods for the simplification and rationalization of work, and establishing new business organizations, are all included in the category of innovation. In simple terms, any of the various ways of doing things differently that occur in the process of capitalist evolution are innovations.

Innovation is not synonymous with the term invention. Initially, it is not significant whether the innovation infers scientific novelty. While most innovations can be traced to some theoretical or practical knowledge that has been developed, many innovations have not required expertise of this nature. Invention does not necessarily follow from innovation, and innovation does not necessarily follow from invention.

The making of an innovation and carrying it out are two different phenomena. Innovation and invention may, or may not, be carried out by the same person. Different attributes of the inventor (curiosity and intellect) and the businessperson (goal driven and volitional) result in the latter turning invention into an innovation. Once innovation is divorced from invention, innovation becomes a direct force for internal change. Creating new factors of production or turning existing factors of production towards new purposes is a purely economic process.

The term *economic evolution* is used by Schumpeter to mean changes in the economic process occurring as a result of innovation. It is used rather than the word *progress* because the latter has undesirable associations suggested by it. Innovation, Schumpeter proposes, is at the center of almost all economic phenomena, difficulties, and issues in the economic life of capitalism. Capitalism is very sensitive to disturbances. Almost all economic phenomena, difficulties, and issues in the economic life of capitalism including its sensitivity would be eliminated if productive resources flowed in either unvarying or constantly increasing quantities. These flows would take place toward essentially the same goals and basically through the same channels.

Innovation and the Production Function

The production function is defined by the way in which the quantity of product varies if quantities of factors vary. If, instead of varying the quantities of factors, the form of the function is varied, innovation results. This, however, limits one to innovation that is made up of the same kind of product and the same means of production used before. To avoid this, innovation will be defined as establishing a new production function. This includes the situation of a new good or service, a new form of organization (such as that created by a merger), the opening of new markets, and the creation of new economic resources. Simply, it consists of new innovations and new combinations.

Schumpeter also defines innovation referring to money cost. In the absence of innovation, the goods and services of businesses and industrial firms, exhibit constant factor prices increasing at a constant rate in relation to their output. If, at any time, a given quantity of output costs less than a previously equivalent or smaller quantity did and prices of factors have not declined, there must have been innovation taking place. Each time there is innovation, the old total or marginal cost curve is nullified and replaced by a new one.

If prices of factors vary independently of the action of the business or industry, the effect on its cost curves (total, average, and marginal) is that the old cost curves become nullified and replaced by new ones. It is apparent that a theoretical cost curve cannot be prepared (as contrasted to an historical one) that would refer to a wage rate at one time and at another time refer to a different wage rate.

Dominating the portrayal of capitalist life and more than anything else affecting our view of the prevalence of declining cost, disequilibria, bare-knuckles competition, and other consequences of an active economy is innovation. Innovation means that the system is constantly adjusting to new production functions and the associated shifting of cost curves. The view that businesses and industries moving in steps of declining costs are at the center of variations in business and industrial life is correct. This is characteristic of innovation, because business and industry are moving along intervals of decreasing costs, changing the existing business and industrial structure, and perhaps sometimes heading towards monopoly. These conditions are those that have established new production functions and find business and industry trying to establish their market.

Schumpeter observes that major and sometimes minor innovations involve the construction of new plant and equipment. This may take the form of the rebuilding old plant and equipment for purposes of reducing costs and reducing rebuilding time. The reverse—that every new plant involves an innovation—is not true. There may be no relation of this circumstance to innovation. The addition to plant may occur to accommodate increased demand. The relative importance of such circumstances varies to the point it is difficult to determine. This collection of factual statements is the source of one of the most difficult processes of estimation that is faced.

In an economy that is undergoing consistent evolution, it is likely not far from the truth to declare that almost all new plant that is being constructed beyond replacement, and much of what is being built by way of replacement represent some type of innovation or are caused by circumstances that relate to some type of innovation. Most new businesses or industries are founded with an idea or a definite goal in mind. They become obsolete when the idea or goal has been achieved. If they do not become obsolete, their newness fades. Fundamentally, this is why businesses and firms do not last forever. Many firms are born that cannot survive. Their inability to find a market for their goods or services, poor management, poor location, and other reasons testify to the failure of businesses and industries to thrive. Businesses and industries that had once come into being with original innovations die a natural death, as they are unable to keep up the pace of innovation they find themselves in. This describes the process of capitalist evolution. There is a constant rise and decline of businesses and industries—a central condition of a capitalist economy.

One of the conditions of innovation is the entrance into the ranks of leadership. The introduction of new leadership into a business or industry may involve the formation of new firms, the transformation of old businesses or industries, or increased competition for old businesses and industries. In the case of large businesses, a constantly changing environment occurs as the firm goes from one innovation to the next.

The ability of large firms to adapt to and create innovations is also part of the process of capitalist evolution. This form of organization does not, however, occur throughout the economic system. Large firms respond to the innovations of other firms. They do so only by going through a process of institutional change. Nonetheless, large firms, over time, ascend and decline on the basis of their ability to innovate.

Innovations may become possible in a continuous process of absorption by the building of new organizations and the alteration of older ones. However, disequilibria are observed that suggest the process of innovation, while occurring with some consistency, is not a regular, continuous one. The process of absorbing innovation is a distinct and stressful operation. This occurs as some businesses and industries implement innovations and then respond by acting along new cost curves. Other firms are unable to adapt themselves and respond by dying.

It is well known that doing something new can be much more difficult than doing something that is routine. Choosing one over the other involves taking a course that varies in both degree and quality. The reasons for this can be grouped into three classes. Initially, the environment resists, as it is familiar with the repeating routine acts. Secondly, the repetition of the production of routine goods and services draws lenders more readily. Labor for routine purposes is readily available. Customers more readily purchase what they are familiar with. Thirdly, many persons feel reluctant to deviate from a routine course of action. This applies to both the production and consumption processes.

When a new production function has been established and its major issues solved, persons adapt to the newness and even make an effort to improve on it. There often is an effort to copy the innovation, improve upon it, and go into competition with it.

Innovations overlap each other. They are fluctuations and irregularities in the innovation process. Innovations are not randomly distributed. Innovations vary in their impact. Some innovations, by their nature, effect major changes as well as secondary and tertiary innovations. For example, a railroad through an area not yet served by rail fosters a period of adjustment and adaptation. Another such "big" innovation is that of electricity. Such innovations spread a wave of change and perhaps innovation throughout the economic system. Economic evolution is discontinuous, irregular, and sometimes overlapping by its nature. This can only be explained by facts not included in a pure model.

Enterprise

The term *enterprise* used to include all actions that involve implementing innovations. The individuals who carry out innovations are termed *entrepreneurs*. No one is ever an entrepreneur one hundred percent of the time. For example, the entrepreneur, as his or her business or industrial firm ages, becomes more of a manager in the process. The entrepreneur may or may not be the originator of the goods, services, or processes developed by him or her. The individual or group furnishing capital, again, may or may not be the entrepreneur(s).

The exposure to risk is no part of the entrepreneur's function. Only those entrepreneurs who are also capitalists are exposed to risk. Secondly, entrepreneurs are not a social class. If they are successful, entrepreneurs or their progeny may become part of the capitalist class. Entrepreneurs originate from all classes. Entrepreneurs originate from the aristocracy, artisan class, farmers, peasants, professionals, and working class. While entrepreneurs may become stockholders in firms, entrepreneurs are not made by the holding of stock. Stockholders are defined by the fact that they are capitalists or creditors who invest, at some level of risk, for the option to participate in the receipt of profits.

The entrepreneur, in a perfectly competitive economy, produces a good or service already in common use at a lower per unit cost than his or her competitors by using a lesser amount of the factors of production. The entrepreneur's receipts exceed his or her cost, the difference being termed *profit*. Profit is the inducement available to entrepreneurs in a capitalist economy. Profit will likely decline in the subsequent process of competition and economic adaptation. Profit represents a net gain not absorbed by the cost factors of production. For profits to occur, it is important that competition at lower costs not be present instantaneously. A period in which investment is regained and exceeded is necessary for profit to occur.

Innovation and enterprise are among the most important sources of speculation that produces windfall gains and losses. The majority of private fortunes in capitalist society are a consequence of the innovation process. Innovation is an important source of business and industrial advance.

Profits bring about the course of competition between businesses and industries. Many enterprises are placed on the defensive as soon as they exist. Profits may be an ominous threat to the structure of a business or industry. Entire sectors may be affected. Unemployment may be a consequence. In some instances, business or industry may make efforts to sabotage an innovation or the process of innovation. In business and industry, taken in its entirety, there is always the old sphere warring with the innovating portion of the economy. Business and industrial strategies develop around this conflagration. For example, price wars, mergers, and attempts to discredit competitors are all a part of competitive strategy.

Economic Evolution, Money, Banking, and Interest

Many of those individuals desiring to be entrepreneurs do not own in part or in whole the where-with-all they require to implement their plans. As a consequence, entrepreneurs borrow most of the funds they need for constructing and operating their businesses and industries. Credit creation by banks makes this possible. This relationship is essential to the operation and understanding of the capitalist economy.

Credit creation is the "monetary complement of innovation" (Schumpeter, 1939, p. 111). This credit may come from unemployed resources or may be withdrawn from other uses. The implementing of an innovation requires an increase in existing factors of production or a transfer of current factors from old to new uses. An innovation financed by savings involves a transfer of funds from one purpose to another. If credit creation is used to sponsor innovation, the credited funds may take away from the funds available to "old" businesses and industries.

In a capitalist economy, banks perform the role of approving or disapproving of innovations and the corresponding credit required to carry them out. These banks are a means by which money retention, money processing, and payment of funds to businesses and industries may take place. In the United States, there are two general types of banks. There are *member banks*, and *bankers' banks* or *central banks*. Member banks keep the accounts of businesses, industries, and households. Bankers' banks or central banks keep the accounts of, and create balances, for member banks. What is of import to business and industry is that member banks create credit.

Not to be forgotten is the role played by government. Enterprise may also be financed through the issuance of bonds or government fiat. The theory of credit creation as well as the theory of saving depends on the purposes for which the credit or saving is formed. The gathering aspect of credit or saving creation is secondary to the purpose for which it has been done.

Financing of enterprises assumes a priority among the banking community. Lending and the formation of a means of payment are important components of the economic process. The capitalist process would be incomplete without them. Sometimes, loans to entrepreneurs are only partially repaid. Often, part or all of a loan may be renewed in such a way as to make the method of payment indefinite or lasting. Innovation causes disequilibria, such that other firms may be required to make investments that cannot be financed out of ongoing receipts. These businesses and industries become borrowers also. As economic evolution is in full process, the majority of bank credit becomes assigned to current business, losing touch with innovation or the adaptive processes associated with other innovations. If we include consumers' borrowing and saving in addition to credit allotted to ongoing business and industrial practices, the bulk of a bank's business is devoted to financing the production of goods and services and the lending of funds to brokers such that the funding of credit for the purpose of innovation is in danger of becoming secondary. Nonetheless, the purchase of bonds (as assets) is a transaction that banks can initiate such that they are less dependent on the vagaries of their customer base and the demand for funds.

Interest is defined by Schumpeter as a premium on present over future means of payment. Interest, or more precisely stated, capital plus interest is the amount paid by borrowers for permission to acquire goods and services (that is, without having provided other goods and services to the stream of the marketplace). For interest to occur, some individuals or households need to estimate a present dollar more highly than a future dollar. Individuals, households, and business will pay a positive amount of interest if a current amount can be used so it will yield a larger amount in the future. A loan for zero interest is the best of all possible worlds, as it will exactly cover current costs.

In business and industry, innovation supports the rubric of interest. The profit it can yield to an entrepreneur is sufficient reason for him or her to pay interest. It is a method for obtaining current dollars as a means for receiving more dollars in the future. Borrowing by the entrepreneur is one way of obtaining those present dollars. Not as well accepted is the statement that entrepreneurs' profits and associated gains, which occur in disequilibria brought about by innovation, are the predominant form of interest payments and the only "cause" of the fact that positive interest rates dominate in capitalist society.

Schumpeter's thesis that the capitalist class thrives on a return that derives from innovation or processes derived from innovation (this does not include the financing of consumption). Schumpeter maintains that while it may be objected that innovation in production and commerce is the only cause of interest, it cannot be denied that innovation is a "cause" of interest in the absence of other reasons. The importance attached to present balances follows from the model of capitalist evolution proposed by Schumpeter.

Interest is, essentially, a monetary phenomenon. It is a payment for balances used to purchase goods and services, not the goods or services themselves. Profits are, for Schumpeter's purposes, importantly a temporary occurrence. They do not remain permanent because of the impermanence of production and trade. This impermanence is reason for the lender to shift his or her resources from one opportunity to another. Each opportunity may vary in duration from the next. Some innovations are more lasting than others and may proceed for generations.

Interest and deviation from it serve as a central function and condition in a capitalist economy. It is a sort of a *coefficient of tension*. The premium on present balances contrasted with future balances is established by borrowers. This includes governments, business and industrial firms, and banks and their branches. Behind this system of relationships is the Central Bank making up the transactions between banker's banks and their banking customers. The Central Bank influences the open market. The other banks indirectly influence the money market itself.

A monetary theory of interest parallels a monetary theory of capital. It is an accounting concept. The resources of a business or industry are measured in monetary terms. These two monetary concepts provide for their introduction into economic theory. The monetary theory of capital understands goods and services not as that, but as monetary balances. The increase or decrease of goods and services is not the same as the monetary theory of capital.

Economic Evolution

Schumpeter finds it useful to outline the basics of his framework for his theory of *economic evolution*. Entrepreneurs prepare plans for innovations with expectations for profits. They work with the obstacles to accomplishing an unfamiliar task or set of tasks. An individual or group of individuals determines they will produce a new or improved consumer good or service. The individual or group borrows the necessary funds from a bank. A new business or industry is formed, the construction necessary for producing the goods and services is completed, and orders for new equipment and materials are made. The entrepreneur(s) use the funds on loan from a bank to provide the necessary assets to purchase goods, supplies, and services necessary to put the innovation into practice.

Other entrepreneurs come after the initial entrepreneur in increasing number. The process of innovation becomes smoothed by the experiences and overcoming obstacles that faced the initial innovator(s). A successful innovation paves the way for other innovations in the sector it belongs to. Innovations may be copied in whole or in part, depending on patent limitations. The consequences of an innovation are experienced throughout the economic system. In addition, the amount of expenditure on innovation multiplied by the velocity level in the previous equilibrium provides a rough approximation of the amount the volume of payments will be increased by innovation expenditures alone. The entrepreneurial process spreads over the entire economic system. Its spread is accompanied by changing values that disrupt the equilibrium that existed previously. With these changes, a new business situation comes into being.

The new goods and services become part of the market. If everything fulfills expectations, the goods and services are consumed at the prices anticipated by the seller. Assuming this, then it is further assumed that the new business or industry will continue to produce an ongoing amount of goods onto the market without a change in the producer's production function. The receipts of the producers from goods and services sold will flow to them and their accounts. This will occur at a rate that allows them to repay the debt acquired (plus interest) in the construction of the plant, equipment, and initial operating expenses plus some additional profit.

This leaves the entrepreneur(s) with no debt (unless he or she has acquired additional debt), possessing plant and equipment debt free, and a balance that can be employed by the entrepreneur as profit or as "working capital." If the same conditions apply to other entrepreneurs then these other entrepreneurs who have shown foresight arrive at an operation state. These firms start putting their goods on the market of consumers' goods and services, thereby increasing the total amount of consumers' goods.

These new goods and services come into an economic world at a rate that exceeds the capacity of the market to absorb them. Initially, the first entrepreneur's production will not be sufficient to alter the business situation. There may be, however, a new component of competition introduced. As the process of economic evolution proceeds, it gathers momentum. Changes steadily gain in magnitude and importance, causing disequilibrium. A process of adaptation begins.

Old firms or those unable to adapt to ongoing changes in their business or industrial sector become affected by the new disequilibrium. Some firms adapt to the new methods, goods, or services. These firms are faced with new opportunities for expansion and change. They may move into them new economic space created by innovation. Other old firms are unable to adapt to the changing business environment. In some instances, these old firms begin to decline economically and face extinction. Others may contract and become insignificant in competition with the innovators. Aggregate analysis in this instance does not tell the entire story. It is necessary to disaggregate the occurring process and transition of firms to understand what is occurring.

Schumpeter poses the question, can the innovation mechanism, previously described, go on forever on a plateau of prosperity or come to a stop from reasons inherent in the innovation process as well as by the nature of its own effects and the business situation it creates? Initially, it is important to understand that entrepreneurial activity does not spread itself evenly through businesses and industry. At first, it is focused on a particular good or service. Its potential is definitely limited. Innovation's results act on specific prices and thus establish constraints on how those prices may change. As more and more firms take up production, the downward pressure on price is anticipated by the original firm that implemented the innovation of the good or service. Similarly, those firms who entered the market to produce the innovation later survive if they anticipate the possibilities left to them. As profits diminish, the innovative impulse will be suppressed.

As entrepreneurial activity upsets the equilibrium of the economic system and the release of new products of innovation brings about more widespread disequilibrium, for a period of time there are variations and repeated efforts at adaptation to changing circumstances. The risk of failure is increased, and the possibility of calculating costs and receipts becomes complicated. For further innovations to be carried out, it is necessary for an equilibrium zone to be established. Entrepreneurial activity declines until it ceases entirely.

The economic system responds to entrepreneurial activity. It adapts to new goods and services created, including the elimination of those lines of products and firms incapable of adaptation. Reorganization of economic life takes place. Debt is liquidated. The system of prices and economic values may be adjusted. The sequence of events leads to a new neighborhood of equilibrium. As the entrepreneurial drive ceases acting, and moves away from its previous equilibrium neighborhood, the economic system initiates a movement which, barring external disturbances, is destined to arrive at a new equilibrium zone. The process occurs over time and may exhibit variations and movements backwards. Nevertheless, a new equilibrium neighborhood will be arrived at.

It is a distance from this conceptual framework to the elaboration of historical fact. There are numerous layers of accidental, "external," secondary, and incidental facts and reactions, as well as their interaction, that cover the underlying infrastructure of economic life. In some cases, the infrastructure may be undetectable. Schumpeter refers to this construct as the "Pure Model or First Approximation."

Schumpeter views the process of economic evolution as proceeding in units separated by equilibrium neighborhoods. Each of these cycles consists of two phases. In the first phase, entrepreneurial activity causes the economy to move away from the equilibrium state. The second phase moves toward another equilibrium position. This sequence of events or fluctuations has come to be known as a business cycle. The presentation of business cycles can be done diagrammatically represented by a rough wavelike movement.

It is not unreasonable to state that economic evolution destabilizes the economic world, and it does so in a process that is characterized as cyclical. A theory of economic evolution consisting of external factors in addition to innovations is clearly more acceptable than a theory providing for no cycles but consisting of an undisturbed stationary or increasing flow. It is because economic life is in a state of constant flux that the business cycle exists. Many who have studied the business cycle are persuaded by the logic of one cycle producing the next (see Mitchell, 1927). Thus, Schumpeter uses the theory of economic evolution as a process for analyzing the cyclical nature of historical and statistical data.

Schumpeter points out that the analytical scheme applied is not included in the monetary theory of business cycles. It should be realized the logic of the cyclical process of economic evolution is independent of monetary phenomena.

In capitalist society, the results of innovation are felt while recessive symptoms are happening. It is, likewise, a time in which secondary effects occur. New methods are being replicated and improved; adaptation to the processes used in innovation occurs; adaptation to the introduction of new goods and services takes place; secondary innovations may occur; new investment opportunities occur; other businesses and industries respond by adapting their technological, business, and industrial processes; and, firms that cannot adapt disappear.

The working of Schumpeter's model of economic evolution does not require that it be periodic. There is no cycle or rhythm that is present. There is a process that produces systematic alternating periods of depression and prosperity. These phases occur through the working of the agency of a "cause" or "force." The length or duration of the units in the process of economic evolution varies with the nature of a specific innovation.

Schumpeter maintains that any crisis that has ever come has can be explained by the objective facts of the situation. In the author's view Schumpeter disregards conditions or circumstances that are hidden from view or cannot be analyzed (such as population sex, subgroup characteristic, and income differences. More will be addressed to these circumstances in Part II.

All prosperities restricted to primary processes, bring about a liquidation period that eliminates out-of-date firms. In addition, a difficult process of price, quantity, and value readjustment occurs. Even in the secondary wave, there are unsuccessful enterprises that will not survive the effects of the recession. These enterprises show losses when prices decline. A portion of the debt structure will collapse. Efforts made by businesses, industries, and households to repay loans are made more difficult by banks' efforts to call them in. The banks' interests are in improving liquidity. Its effect is to move debtors towards insolvency. Freezing of credit, deposit shrinkage, and more follow. The danger in such situations is that pessimistic expectations may dominate and take a causal role.

Downward trending of values and the declining of the operations cause firms to move below what might be their equilibrium amounts. In recession, a process is in operation to draw the economic system towards equilibrium. Nonetheless, the economic system moves away from the neighborhood of equilibrium under a different set of factors. This phase is termed by Schumpeter as Depression. After the depression has run its course, the system gradually moves back towards a neighborhood of equilibrium. This is termed by Schumpeter a "Recovery or Revival." The economy expands up to the equilibrium amounts until the equilibrium zone is established. Surplus gains occur or losses are limited. Some losses may have been equivalent to the losses at the level of the trough. A pattern is produced, and the economy becomes inhabited by businesses and firms that can obtain adequate financial support.

If recession moves into depression, there is some difficulty in explaining depression. Depression's occurrence is a matter of fact and may well occur as a result of accidental circumstances. Furthermore, at this time, there are no theoretical expectations that lead to an understanding of why depressions occur.

It is now appropriate to address the question of whether or not the economic system stops once it has entered into a negative phase and moves back towards a more positive phase. Schumpeter's analysis causes him to agree with those who believe the economy has "recuperative" powers. Yet, Schumpeter questions whether a Depression may have a manner of feeding upon itself indefinitely. It is clear that firms will act in such a way as to take steps towards equilibrium and recovery. Eventually, through their actions, they will acquire a new equilibrium or equilibrium neighborhood. The possibility that firms will find themselves in lower-level equilibria from which they will not move cannot be expected to be a system-wide phenomenon.

The question reduces to the issue of whether the depressive process stops itself before total collapse. It can be shown that the stress of the downward trend produces reactions by businesses and industries that tend to halt the trend. The downward trend is marked by a number of distressing individual events. Bankruptcies, the decline of individual markets, and business or industry shutdowns are examples. These events are accompanied by other distressing events. It is readily apparent that each of these events loses its momentum, as a process of "dilution" or "diffusion" takes place. Individual contractions of output foster other incidences of contraction all around. This decline occurs at a decreasing rate. The impact of declines of individual firms gradually decreases. The decline process, initially, is a movement away from equilibrium.

A depression, Schumpeter stresses, is unlike a recession and is a pathological process. The argument that depressions will find a "natural end" does not constitute an argument for having things take their course. Trust in the restoring forces of nature should not be used to reduce the need for government action, whatever one may think of that argument.

The theory of the lower (recession) and upper (prosperity) turning points remains. Schumpeter maintains the division of the units of a cyclical process into two or four phases is more than a matter of descriptive convenience. Each phase is a distinct phenomenon. The phenomenon can be understood by starting in the neighborhood of equilibrium occurring before prosperity, and finish with the neighborhood of equilibrium following recovery. The analysis of this cycle from peak to peak or trough may be unreliable. Analyzing in this manner causes the fundamental distinction between revival and prosperity to be lost. The difference in the two is the propelling factor. In the search for the causes of revival, many analysts look to the elimination of deviations that exist. They examine low stock prices, unemployed labor, unused plant and equipment, and low levels of credit. Specifically, these analysts find nothing that looks like innovation. Thus, they arrive at the conclusion that innovation has nothing to do with encouraging prosperity.

It is important to take into consideration that each neighborhood of equilibrium contains undigested components of previous depressions and prosperities. There is nothing in this to contradict the model of economic evolution. These conditions do, however, complicate the business and industrial situations dealt with. Evolution, in the sense used by Schumpeter, is one of the most powerful influences in developing imperfections throughout the economic system. It is appropriate to assume that both competition and equilibrium are imperfect from the beginning. Imperfections of both competition and equilibrium, and external disruptions may well explain the existence of unemployed resources separate from the cyclical process of economic evolution.

Schumpeter, in his work, advocates a single sequence of cycles, barring the possibility of external disturbances. Each of these cycles is of the same type as its predecessors and successors. Each cycle may be thought of as restricted or drawn out in duration, increased or diminished in amplitude by its historic circumstances, and internally irregular. There are reasons to anticipate that the process of economic evolution should give origins to only one wavelike movement. In fact, there are reasons to expect that economic evolution will foster multiple wavelike movements that will interfere with each other in the evolutionary process. Because of this, it is more acceptable to propose the existence of numerous fluctuations of different length and intensity that are superimposed on one another. The positing of a single cycle theory has created more difficulties than it has been proposed to solve.

The development of cycles of different lengths was accomplished by Kitchin, Jugular, and Kondratieff. Schumpeter advanced a model in which there were four major superimposed cycles within each other. These were a Kondratieff cycle (54 years), a Juglar cycle (9) years, and a Kitchin cycle (4 years). There is important disagreement in the community of economists about the existence of these cycles and what they represent. The analysis of these cycles yields irregularities in the cycles and large standard deviations.

One can only conclude that Schumpeter was either operating with a different definition of business cycles, or that his use of business cycles cannot be empirically verified. The author posits after reviewing Schumpeter's material on business cycles that the latter is the case. Schumpeter proceeds through the remainder of his two-volume work, questioning his framework of cycles while hedging about their veracity. Nonetheless, he keeps returning to their use.

While this rejection of the cyclical pattern proposed by Schumpeter may be accurate, it is not reason enough to throw out his general framework of his analysis. The ideas of innovation and economic evolution have an appeal because of their basis in reality. Change, innovation, growth, and decline are persistent features in an economic system as large and advanced as that of the United States.

Schumpeter's View of the Depression Era

The occurrence of the Great Depression era was prolonged and severe. Schumpeter maintains that the framework he has presented can be used to provide an analytic model for some of the period from 1929 to the summer of 1938. He also asks how must other factors, internal and external, new or old, be relied upon for explanation? No claims, he emphasizes again, are made for the three-cycle schema other than it is a useful descriptive or illustrative tool.

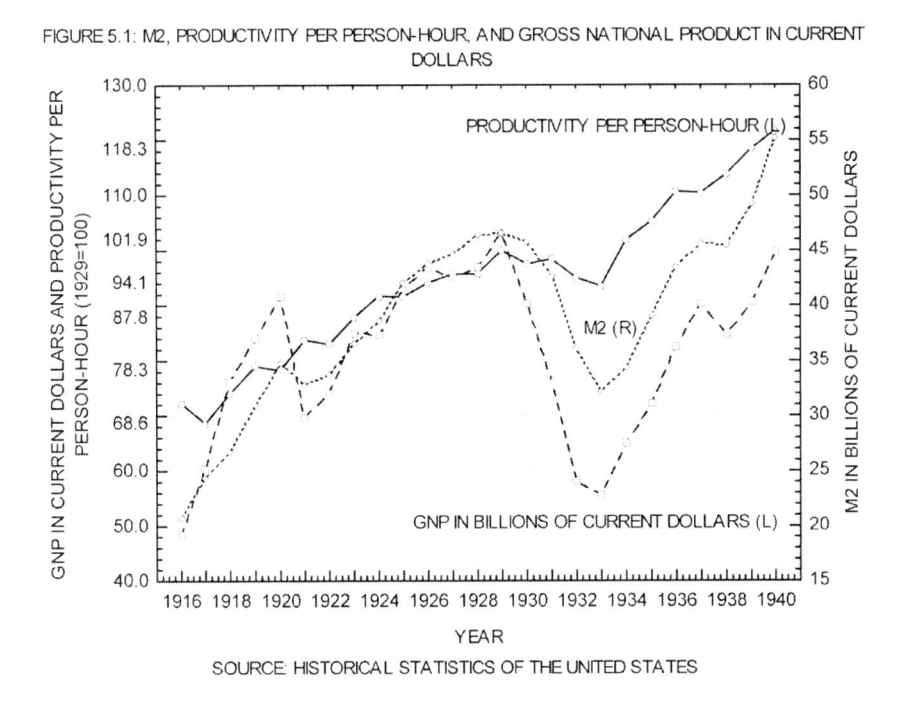

FIGURE 5.1: M2, PRODUCTIVITY PER PERSON-HOUR, AND GROSS NATIONAL PRODUCT IN CURRENT DOLLARS

SOURCE: HISTORICAL STATISTICS OF THE UNITED STATES

Source: GNP Part 1, p. 224, col. 1; PPPH Part 2, p. 948, col. 4; M_2 Part 2, p.992, col.415

Schumpeter's efforts to apply his cyclical model bears no resemblance to the levels of productivity characterized in Table 5.2. The data from Figure 5.1 follow in Table 5.2. Productivity declined in the 1929-1933 period. It is unlikely that this can be attributed to anything other than the lack of demand and the lack of demand for money. Monetarists would focus on the changing money supply during that period. Monetarists can point to a declining money supply (M_2). In contrast, demand proponents can emphasize the decline in GNP due to declining demand (resulting from declining personal income and personal consumption), and as will later be demonstrated, as a radical decline in the number of foreign-born males 15-74 years of age.

Table 5.1: M_2 in Billions of Current Dollars, Gross National Product in Billions of Current Dollars, and Production per Man-hour (in dollars) *

Year	M_2 Supply in Billions	Current GNP in Billions	GNP Productivity per Person Hour
1917	$24.37	60.4	68.6
1918	26.73	76.4	74.1
1919	31.01	84.0	79.0
1920	34.80	91.5	78.3
1921	32.85	69.6	83.8
1922	33.72	74.1	83.0
1923	36.60	85.1	87.8
1924	38.58	84.7	91.7
1925	42.05	93.1	91.6
1926	43.68	97.0	94.1
1927	44.63	94.9	95.7
1928	46.42	97.0	95.7
1929	46.60	103.1	100.0
1930	45.73	90.4	97.5
1931	42.69	75.8	98.4
1932	36.05	58.0	95.0
1933	32.20	55.6	93.5
1934	34.36	65.1	101.9
1935	39.07	72.2	105.3
1936	43.48	82.5	110.7
1937	45.68	90.4	110.5
1938	45.51	84.7	113.8

* Source: Historical Statistics of the United States, See Figure 5.1

From 1929 to 1933, M_2 declined 30.9%. Over the same period, GNP declined 68.7%. The velocity decline in the period was .6 times per year. This level of decline in velocity amounted to a 27.2% change from 1929 to 1933 (see Figure 5.2).

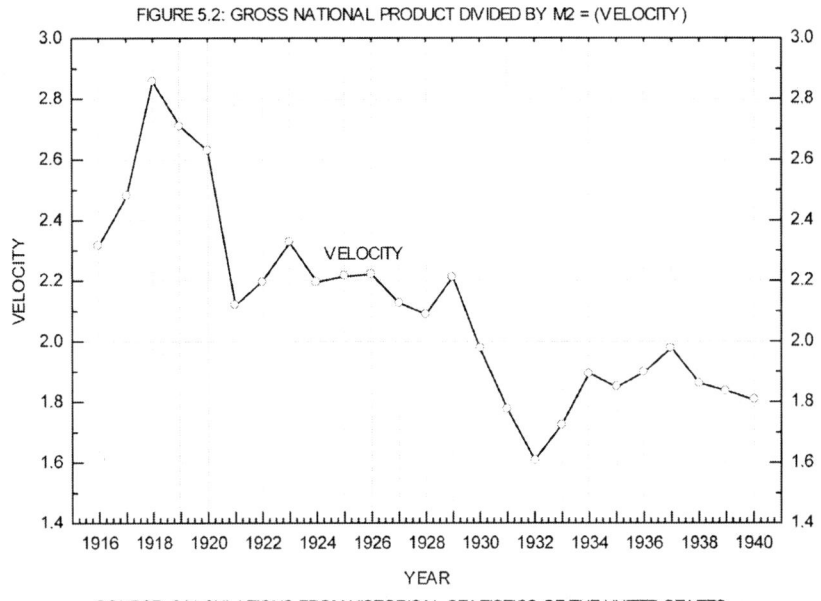

FIGURE 5.2: GROSS NATIONAL PRODUCT DIVIDED BY M2 = (VELOCITY)

Source: Calculated from M_2 and GNP

This macro-analysis raises questions about the validity of any analysis that excludes an examination of the indicators M_2 and GNP. While this does not invalidate Schumpeter's framework of economic evolution, it does point out the need to consider additional factors.

The Depression Era Continued

Capitalist evolution implies disturbance. These disturbances differ in their importance and time span. There are times when the process of eliminating firms and rearrangement occur. Long-range and short-range adjustments are made. Historical observation reveals the extent of change in business and industrial structure. Three-quarters of all firms (farms included) had to adapt. Some did not survive. Was this not the central fact concerning the Depression era? All other factors were merely subordinate or qualifying circumstances.

Schumpeter says of the Depression era, "Capitalism and its civilization may be decaying, shading off into something else, or tottering toward a violent death" (p. 908). This is Schumpeter's view. He qualifies these remarks by saying they only cover the latter half of 1932. The depression era was not symptomatic of a weakening or failure of the capitalist system. It provided proof of the strength of capitalist evolution to which it was a temporary reaction. In fact, it was an occurrence that resembled events that had happened in previous times.

He recalls that in his writing, he never undertook to explain or succeeded in making clear everything about numerous crises or even depressions. The boom and Crash in the United States stock market, though strongly influenced by the process advanced by Schumpeter, is in no way an important part of the process of economic evolution. Schumpeter warns of extending this position too far. Building booms and their decline from about 1928 on as well as the changes of the agricultural sector should be regarded as normal parts of economic evolution.

The debt situation of Americans and the three American bank epidemics were major contributors to the depression era. The accumulated load of debt combined with a fall in price levels were important factors precipitating the Depression. The debt-deflation theory applies to this instance.

The decline of imports was a critical factor. Imports into the United States totaled $4.4 billion in 1929. They declined to approximately $1.3 billion in 1932. This decline of approximately 70.5% again (from the author's perspective) represented a decline in demand.

In the autumn of 1929, businesspersons and forecasters foresaw that there was nothing in store other than a recession with its subsequent recovery. This outlook carried over into the first half of 1930. The latter half of 1930 witnessed a very different scenario. There was widespread liquidation of businesses and industries. Demand deposits did not decline appreciably in the latter half of 1930. An exception prevailed. All other loans declined significantly from 1929 ($41.9 million) to 1933 ($22.3 million) and continued to decrease. The total number of banks declined during that same period by 10,797. Many of the banks were not members of the Federal Reserve System. Excess reserves of member banks were at about $475 million. With the exception of July and August 1930, there was a net influx of gold in all months of that year. The influx of gold plus the issuance of bank notes brought excess reserves of member banks to approximately $475 million.

The number of business failures was above 1929 levels. Business failures were occurring at the rate of 2,000 a month. While high, this was below the monthly failures from October 1921 to June 1922. The question arises: do these facts fit with the theory of economic evolution? Schumpeter maintains that they do.

Schumpeter admits that his methods do not permit him to quantify the expectations for his theory of economic evolution (p. 915). He maintains that the facts of the situation assist him in maintaining that the process of economic evolution was a definite factor. He contends that without his theory, overinvestment and overproduction would be incapable of being explained.

A cautious Schumpeter warns that the possibility that other factors influenced the course of events could possibly result in a spurious conclusion. The effects of measures to affect the circumstances of agriculture failed to modify conditions (foreclosures, the Dust Bowl, low prices) adequately to affect the economic system in its entirety. Secondly, the Smoot-Hawley tariff generated reprisals, which zeroed out its impact.

When it became obvious that business was headed for a recession, the Federal Reserve System resorted to what the public and some economists believed to be the remedy: large scale open market purchases. From October 1929 to August 1931 open market purchases amounted to about 602 million dollars. After August 1931 purchases declined to small amounts. Rediscounts were paid off by member banks while increasing investment. A concern of banks was finding customers for their available funds. Schumpeter maintains that the reserve system, which favored cheap money and expansion, had not been an important factor affecting the business situation.

Severe depression symptoms persisted from 1931 through 1932. In fact, the happenings and circumstance correspond to the expectations of our theory of economic evolution. The deepening of the recession into depression is marked by "normal liquidation" into abnormal liquidation."

The decline is marked by irregularities that are unpredictable and responsive to external factors, and unpredictable incidents.

Contrary to Schumpeter's observation that the Depression era acted as an efficiency expert, the author finds that the period from 1929 through 1933 is a period of lower productivity. Not until 1934 did productivity regain levels above those of 1929.

Gold, U.S. Banking, and the Depression

England suspended gold payments on September 20, 1931. Schumpeter thinks that this event led to a generalized distrust in the currencies of nations. Inflationary possibilities arose in the American position, and the possibility arose of a domestic run on the relatively small amount of "free" gold that was available. Many European banks made efforts to convert holdings of American exchange into gold. This outward flux of gold was stayed by President Hoover's consultation with a French representative. To a large extent, the French had, to a large extent had their currency balances repatriated, so France agreed to stop withdrawals, understanding that their experience with the British would not be repeated.

In February through June 1932, a smaller scale withdrawal of gold occurred. A reversal of the flow of gold stock in the last half of the year found the U.S. stock of gold reserves slightly increased.

Another component of the economy was dramatically impacted as the 1929 through 1933 period progressed. The impact was major. From 1929 through 1933, the number of banks declined from 215,568 to 14,771 (-10,797, or -42%). The second epidemic, which was not as serious as the first epidemic, and the rapid increase in business failures, were not adequately measured by losses suffered by banks and other creditors from bad debts. The drain of payments that were made, the fact that the value of collateral was diminished, and the net worth of many people declined suggested that business and industrial operations would be generally restricted. Declining wealth and decreased employment were staples of the situation. It took time for people, businesses, and industry to realize that they need not be in hope of a speedy recovery.

Soon after the president signed the Glass-Steagall Act (February 1932) the Federal Reserve System, in a large open market operation, purchased $1.11 billion of government bonds from March 1932 to August of the same year. The result of this operation was to effect the reduction of interest rates and cause the buildup of excess reserves. This had the effect of stabilizing the outflow of gold in May and June of that year.

Government Income Generating Expenditures and Gold

The RFC (formed in January 1932) and its extension, the Emergency Relief and Construction Act (ERCA) (July 21, 1932) were the most substantial of the efforts to dampen the effects of the depression. Among the firms it strengthened were banks and trust companies, building and loan associations, insurance companies, mortgage loan companies, and railroads. As of September 30, 1932, the amount advanced was approximately $1.2 billion. One hundred eighty-five million dollars had been repaid by that time. The RFC issued $750 million of 3.5% notes of which $600 million were purchased by the Treasury. The RFC was designed to affect banks and similar financial institutions. The only big business included was the threatened railroads. This avoided additional failures, although there was not much in the way of positive results coming from the funds loaned.

During the Hoover administration, declining land values and farm prices were responsible for foreclosures and an increasing number of forced sales due to tax delinquency. The ERCA authorized funds for relief, and work relief was ineffectual. Thirty-five and one half million dollars were authorized for relief purposes and only $14.2 million were spent as of the end of September 1932. The urging of the President for decreasing public expenditures and increasing taxes sent a mixed message. Promoting fiscal responsibility and balancing budgets were unlikely methods of meeting current crises while proposing balanced budgets in the future. Federal expenditures designated to generate income were an estimated $1.728 billion in calendar year 1931, and $1.646 billion in calendar year 1932. This can be thought of as the most effective part of government policy. It was designed to prevent additional disasters. In spite of this effort, the tide was not turned.

In November 1932, Nevada declared the first of a rash of banking holidays. The increase in banking holidays in January and February of 1933 was stopped by federal legislation on March 9, 1933. By March 5, 1933, there was almost a complete cessation of banking activity. The agricultural mortgage situation was a main cause of this stoppage. A hurricane started in the middle and farther West, moved into the Midwest, and on to the East. Panic occurred. Well-known was the distrust in banks, distrust in the currency, and the widespread withdrawal of deposits. These activities caused banks to withdraw currency from reserve banks. Member banks withdrew over $1.7 billion from February 8, 1933 to March 3, 1933. On March 2nd and March 3rd, Federal Reserve credit increased by $730 million. On the same days in 1933, money in circulation increased by approximately $700 million. Difficulties were accented by the fact that excess gold reserves of federal banks were reduced from $1.1 billion to $400 million. The movement of gold out of the United States was over $270 million in February and March of 1933. Because of the Emergency Banking Act of March 9, 1933 (amended on March 24th), the banking holiday was over, gold certificates and coins in the amount of $600 million were returned to the federal reserve banks by the end of March. Excess gold reserves increased to $1.172 billion due to the reduction of the amount that was to be held to back notes outstanding. The number of (all) banks increased from 14,771 at the end of 1933 to 15,913 at the end of 1934. The consequences of the panic on the American people were dramatic. People were demoralized and saw nothing but continuing disaster in their future. The year 1934 offered some hope of recovery.

Schumpeter continues to apply his tripartite cyclical thesis, as he continues to hold to its veracity and ability to assist in understanding economic evolution. He asserts that potential recovery is consistent with further decline in the dollar amount of business operations. Schumpeter states that the symptoms of economic disease that he wishes to address at this point in his analysis are unemployment and prices. The annual maximum of unemployed occurs in 1933 (12.83 million). Despite this, employment in manufacturing industries started to increase slightly in July 1932. Prices declined by about 24% from 1929 (CPI=51.3) to 1933 (CPI=38.8) (1967=100).

Many manufacturing industries withstood the reduction in prices. Examples were food, house furnishings, leather product industries, paper, petroleum refining, printing, and tobacco. Equipment industries returned to a lesser extent or not at all. Reduction in prices hardly stimulated demand.

Personal income declined from $85.9 billion in 1929 to $47 billion in 1933 or 45%. Personal consumption fell from $77.4 billion in 1929 to $45.9 billion in 1933 or 41% The number of profit-making corporations reporting losses was greater than the number reporting profits in 1930. This statistic showed the ratio of the number of concerns reporting losses to those with net gains continued declining through 1932 (data is not available for 1933). The ratios reported were: year 1931, 284 to 176; and 1932, 366 to 80. Corporate accumulations followed a course of decline also. They were: 1930, minus $4.11 billion; 1931, minus $6.04 billion; 1932, minus $6.55 billion. Business savings reported by the Department of Commerce also showed a decline from 1930 through 1932.

Schumpeter dismisses the fall in hourly wage rates in some industries as a cause of the Depression. He argues this because wage rates did not fall until the contraction was in full swing. Schumpeter thought it difficult to say whether wage rates deepened or weakened the Depression. This was so because the downward shift of the demand curves for labor of individual businesses and industries may have failed to make demand for additional labor necessary. There were likely also situations in which the reduction in wage rates resulted in a decline in employment and output. This observation loses its cogency once recovery begins and demand curves for labor tend to move upwards. Schumpeter concludes that a decline in wage rates stimulated recovery and that a stronger fall in wage rates in the United States would have fostered a stronger recovery.

Recovery

In 1931, recovery policy in the United States was a minor factor. It was of somewhat greater importance in 1932. In 1933 and beyond, recovery policies were important parts of the economic scene. This was so much so that Schumpeter raises the issue of whether there is any purpose in proceeding to using his tripartite cyclical phases, attempting to debate them, or relating actual business situations to the process of economic evolution. From 1933 on, a new economic pattern came into being. The new economic formation calls for a revised economic model, new data, and a new structure. This is particularly true of the investment process.

We are faced with a question of fact, not of principle. The main principle involved is that which rests on the "certainty" that the economic processes of a capitalist society will evolve into a different form of economy. Schumpeter maintains that it need not be asked whether the U.S. economic system could have recovered without government stimulating it out of deep depression. He maintains it did. He argues that it would be contrary to previous experience as well as common sense to argue that a process that had been operating since the sixteenth century up to the end of 1932 would suddenly stop on March 4, 1933. Although an era of social reconstruction occurred, government intervention is labeled an external factor in Schumpeter's model.

The passage of more than eighty acts by the Seventy-third Congress up to the first week in June 1933, as a group, steadied the situation present at that time. Recovery was enhanced by the Emergency Banking Act, which was passed and signed by the end of March 1933. It provided for the reopening of closed banks. This was followed by the Banking Act of June 1933, providing for a strengthening of the Federal Reserve's power over its member banks. Open market operations became centralized, and deposit insurance was introduced. The regulation of credit for speculative purposes was included.

In addition, numerous issues of credit and debt were addressed. Among the most important of these was the Farm Relief Act (May 1933), which provided agricultural credit and addressed the need for refinancing farm property. The Farm Mortgage Corporation Act (January 1934) and the Home Owners' Loan Acts (June 1933 and April 1934) addressed another part of the credit structure that affected a large number of individuals as well as banks.

These and other actions taken by the federal government did not, according to Schumpeter, make recovery possible. He asserts that such actions assisted in providing institutional and other conditions (not stimuli) for the process of recovery to resume soon after the economic failure of in March 1933. The combination of recovery measures had a salutary effect that must be ranked highly. Yet, Schumpeter maintains that the recovery process, without these efforts, would have proceeded stimulated by the factors of his model.

Schumpeter claims that he would arrive at no different conclusion when considering the AAA and the NRA. While a distinction deserves to be made between welfare and prosperity, both Acts encouraged recovery of the usual kind and did not replace the recovery by one that needed to be explained by different principles. These two Acts replaced parts of the capitalist system in a manner that was corrective rather than constructive.

This is essentially true in the case of policies for agricultural adjustment. It dealt with problems that had been a long time in the making. Wholesale farm liquidations would have been an "automatic" way of restoring equilibrium. The alternative to this would be to preserve a large, contented farming population, thereby preserving a permanent disequilibrium. Extensive liquidation of farms that affected business and industrial unemployment (it was, itself, unmanageable) would have had an impact with re-migration to the land. This would have been an "automatic" path to reestablishing equilibrium. It would restore previous processes in the agrarian part of the economy and the relations of the agrarian to the business and industrial sectors. The problem and its solution were straightforward. Refinancing serious farmers subject to foreclosure, nationalizing the marketing of agricultural products, emphasizing the export of agricultural production, and planning production in a rigorous way would all be parts of the adjustment process. United States public opinion and the likelihood of violating the Constitution were in the way of these steps being taken.

An indirect route was taken to address the slumping agricultural sector. The Farm Relief Act (FRA) was passed in May 1930. By increasing farm revenue, it follows that there was some effect on the Depression. Net results were difficult to evaluate. Moreover, it was difficult to assess how much was due to the root idea of the Act, which was to restrict production and reimburse farmers for that which they failed to plant and grow. Programs were addressed and affected by legislation in the areas of tobacco farming, dairy farming, wheat production, and hog farming. Bad harvests of wheat from 1933 through 1936 made the program for wheat unnecessary.

The farming of cotton was not interrupted by poor natural conditions. Reducing acreage for 1934 and 1935 was made more effective by a bad cotton crop in 1933 (approximately 25% was destroyed). Annual prices for cotton in 1934 were approximately twice that in 1932, with the government holding over 3 million bales. Furthermore, participation was encouraged by loans for seed and credit from the Commodity Credit Corporation. Growers' prices (including benefits) increased from $483 million in 1932, to $880 million in 1933, to $893 million in 1934.

Title I of the National Industrial Recovery Act (NIRA) (June, 1933) included codes for fair competition and introduced state-regulated industrial business and industry. The Act assisted in recovery by identifying weak spots in business and industry, stopping additional declines in certain businesses and industries, and correcting disorganized markets. This was especially true in situations that featured overproduction and encouraged underselling obsolete products. In some cases, the Act interfered with the process of business and industrial transformation that was proceeding. There was a failure to see the process of transformation that was going on. In other instances, the Act brought improvement to situations that were disorganized or required a less-extensive measure. The Supreme Court invalidated the Act in June, 1935.

The NRA regulated wages and hours. Schumpeter stated that he thought the entitlement to collective bargaining and the banning of the yellow-dog contract were not sufficient cause to prevent recovery. He did recognize that the effort to raise the wage rate structure had a deleterious effect on employment per unit of production and the expansion of production. The lack of advancement of employment, output, and payrolls gives credence to the presence of this restraining force. Labor became costly relative to real capital. Schumpeter thought the existence of a high-wage rate policy was a definite precursor to high unemployment.

Schumpeter goes on to apply his cyclical theory to an overview of the events of 1932 through 1933. He designates three considerations as sufficient to explain the coming recovery in 1934. First, a potential recovery occurred in 1932. Second he thought the banking collapse in the spring of 1933, conditions in the agrarian sector, governmental indebtedness, and the depressive conditions in individual industries interrupted that incipient recovery. Third, the series of measures inaugurated by the Roosevelt Administration (with the exception of higher wages) were in their entirety remedial in their effect. Schumpeter maintains that these three groups of considerations were adequate to explain the immediate recovery of the capitalist economy in the United States.

What was a "natural" recovery was underway. By this is meant a recovery coming about in a cyclical process as a consequence of its mechanism. Secondly, the process of recovery was "sound". A sound recovery is defined as a recovery that does not contain tendencies to produce similar difficulties as existed previously or introduce new difficulties in place of those that have been overcome. Sound recovery is not necessarily natural. The final effects of NRA were substantially sound. Nonetheless, Schumpeter maintains that the underlying recovery was natural.

The prevailing unemployment makes any implication that the Congress or administration "should" or "could" have not implemented their program of corrective measures unwarranted. According to Schumpeter, recovery started by "itself" in the second and third quarters of 1933. At that time, the new spending program of the administration was still in its incubation stages. The federal income-generating expenditures in 1932 were $1.856 billion. In 1933, these expenditures increased by only $210 million. Thus, Schumpeter maintains, the spending program of the federal government did not have a major impact at the time.

One factor that Schumpeter overlooks is the success of the legislation establishing the Federal Deposit Insurance Corporation (FDIC) in late 1933. It demonstrates the importance of the FDIC to greater confidence in the banking system and the decline in the suspension of banks. Of interest is the peaking of bank suspensions (22,241) in 1931, followed by a minor peak in 1933 (8,000), and a reduction in 1934 (114) (see Figure 5.3 and Table 5.3). The corresponding dollar amounts of deposits suspended were $3.380 billion in 1931, $7.193 billion in 1933 and $73.87 million in 1934.

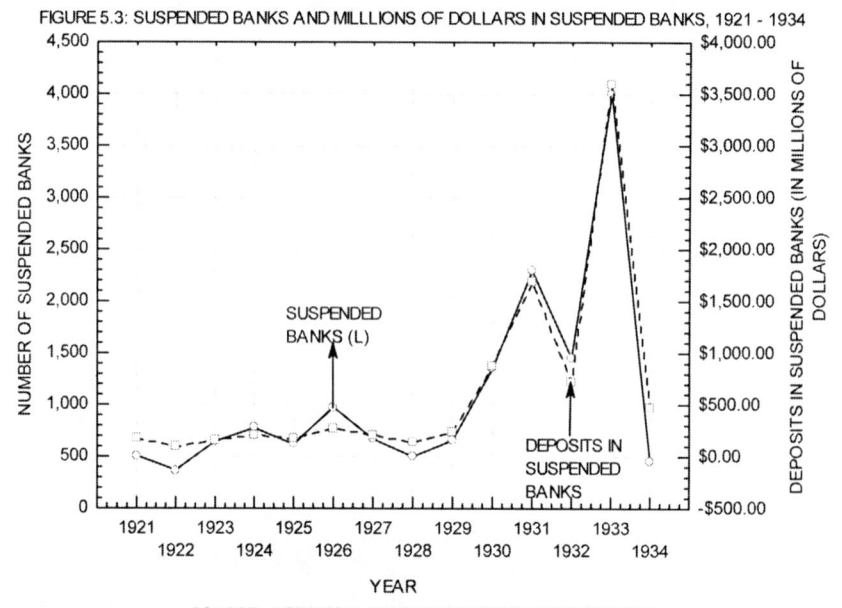

FIGURE 5.3: SUSPENDED BANKS AND MILLLIONS OF DOLLARS IN SUSPENDED BANKS, 1921 - 1934

SOURCE: HISTORICAL STATISTICS OF THE UNITED STATES

Source: Table 5.2

Table 5.2: Number of Banks Suspended and Dollars of Suspended Deposits

Year	Number of Suspended Banks	Amount of Suspended Deposits
1929	659	$ 231,000,000
1930	1,352	869,692,000
1931	2,294	1,691,000,000
1932	1,456	725,000,000
1933	4,004	3,601,000,000
1934-1940	448	477,000,000

Source: Historical Statistics of the United States, Part II, p. 1038, cols. 741 and 748

Table 5.3: Bond Rates 1932-1935, Wholesale Price Index, and Consumer Price Index With Annual Percent Change: 1929-1940

Year	Bond Rates*	Wholesale Price Index**	Annual Percent Change	Consumer Price Index***	Annual Percent Change
1929		95.3		51.3	
1930		86.4	-10.3%	50.0	-2.6%
1931		73.0	-18.4	45.6	-9.6
1932	6.27%	64.8	-12.7	40.9	-11.5
1933	5.92	65.9	+1.7	38.8	-5.4
1934	4.86	74.9	+12.0	40.1	+3.2
1935	4.78	80.0	+6.4	41.1	+2.4
1936		80.8	+1.0	41.5	+1.0
1937		86.3	-6.4	43.0	+3.5
1938		78.6	-9.8	42.2	-1.9
1939		77.1	-1.9	41.6	-1.4
1940		78.6	+1.9	42.0	+1.0

* Schumpeter. p. 1008
** Historical Statistics of the United States, p. 200, col. 40, (1926 = 100)
*** p. 210, col. 135, (1967 = 100)

The effectiveness of the FDIC needs not be debated. Its role in making deposits more secure provided a solid financial base, which was reassuring to the clientele of banks as well as those who had withdrawn funds. It was one of a number of conditions that were important to a slow recovery in the era of the Great Depression.

Gold Again, Devaluation, and Banks

After the banking holiday in March 1933, reopening banks were restricted from paying out gold (except by special permission of the Department of the Treasury) and using gold to redeem notes terminated on March 4, 1933. Despite this, an insignificant decline in the value of the dollar occurred. After the restriction on gold flows, dollars traded internationally in a range of between 8% and 12% of its former price. Persons speculating on the international market did not believe the dollar would depreciate substantially. The dollar was not under pressure from abroad or at home in either a short-run or long-term sense. Within three weeks of the banking holiday, over 50% of the loans to reserve banks to meet the loss of gold and withdrawal of notes had been repaid. The losses suffered by reserve banks were made good.

The dollar fell with a hesitation step; and in the autumn of 1933, recovery slowed. The RFC was not successful despite its decision to lend banks up to $1 billion for the purposes of re-lending. The program failed. There were numbers of individuals, businesses, and industrialists proposing that inflationary policies be initiated. The administration reacted by proposing a method of bringing the dollar down without inflation. It was proposed that gold buying would bring the value of the dollar down. Following the presidential proclamation of January 1934, the price of gold went from $20.67 to $35. Approximately $381 million of gold was imported into the United States in February. The value of the monetary stock stood at $7.03 billion. In February, 1934, the influx of gold brought the excess reserves of member banks to $1 billion. The gold dollar was devalued to 59.06% of its former weight. The Treasury stood to gain a paper profit as it had formerly valued its gold holdings at $20.67 an ounce. In the spring of 1935, this was a primary factor in the adoption of very low lending rates.

Schumpeter stated that the meaning of devaluation must be considered in relation to the level of public expenditure. The meaning of devaluation consists primarily in implementing the latter. The belief that redefining the gold content of the dollar would proportionately change the price level was naïve. That view was a survivor from the days of the commodity theory of money. Devaluation can only act on prices if it brings about or makes possible an increase in expenditure. Net national income increased by approximately $8.6 billion in 1934, $5.2 billion in 1935, and $8.8 billion in 1936. Federal expenditures generated $1.856 billion of income in 1933, $3.238 billion in 1935, and $3.154 billion in 1935.

The influx of gold occurring was so rapid in the November 1934-September 1936 period it precipitated action by the Department of Treasury and the Federal Reserve. U.S. devaluation had resulted in the influx of gold. The United States was unwilling to retreat from devaluation. Although the gold influx in 1934 was significant, the gold influx in 1935 amounted to $1.75 billion (over 50% from France). Near the end of September 1936, the U.S. Department of Treasury entered into an agreement with England and France, which helped control the movements of gold temporarily. In December 1936, the United States initiated a gold sterilization plan that proved to be a successful method of acquiring and retaining gold. It avoided, up to September 1937, negative effects of the gold influx on bank reserves and deposits by increasing the monetary gold stock and treasury cash going into an inactive gold account. Of the approximately $1.9 billion increase in monetary gold stock, $350 million was in circulation and $1.5 billion was in members' reserves. One billion dollars of this was in excess reserves. In 1937, the gold influx increased. Monetary gold stock increased by 1.5 billion dollars. In November and December of 1937, there was some reversal in the flow of gold stock.

The Department of the Treasury desterilized 300 million dollars of gold in September, 1937 and then reduced the sterilization decree to a shadow of its former self in February 1938. The decree was reversed by the release of 1.4 billion dollars of gold in the inactive account in April 1938. This was accomplished by transferring that amount to Treasury accounts with Federal Reserve banks. Margin requirements for security transactions were decreased in November 1937. The following April, reserve requirements were reduced by 13.5 percent. Reserve member banks' reserves rose by the end of 1937 due to the release of 300 million dollars of gold. In addition to this, there were monetary silver, open market operations, and later additional purchases of gold to add to reserves.

Federal Expenditures

Initially, Schumpeter maintained that federal income may be affected by his underlying cyclical process. Any expectation about its impact needs to be increased not decreased. Government expenditures will improve any business situation if they increase. Public expenditure may increase overall expenditures by building up depleted balances or repaying debts. Even a decline in demand deposits may be a favorable sign. A moderate amount of freeing up resources remains for the early part of the recovery phase.

Secondly, there are direct consequences of expenditure of government funds. Unemployment expenditures, the filling of government orders by business or industry, and an employee of that firm expending his or her wages are all examples of direct expenditures. Thirdly, firms, particularly those experiencing recovery, respond not only directly to government purchases or orders but also indirectly by expanding operations in response to anticipated increases orders or purchases. During the early phases of recovery, such new investment cannot be expected to be significant. Lastly, it is possible and even likely there will be "ulterior" effects on the economy. Federal income generation also can be expected to provide a stimulus directly and indirectly to consumers' credit. The general feeling that a floor (FDIC) has been provided, relief in the debt structure, stabilizing of prices, and improvements affected by government expenditures provides an all around impact. While Schumpeter does not deny the reality of such effects, he maintains they should not be used solely and uncritically.

Schumpeter's analysis, while it does not tend to undercut the potential of pump priming, it may, if sufficiently persistent, alter depression into a condition identified by all the surface characteristics of prosperity. Schumpeter concludes that other considerations weakened the direct and ulterior effects of pump priming. Many economists accept the introduction of direct and ulterior purchasing power effects as adequate proof that there could not be contributions from the economic process itself. Neither would these economists expect there to be a negative impact of pump priming.

If government expenditures increase business and industrial deposits as is expected, the government expenditures will support the subsequent transactions of the recipient firms, whether such expenditures are caused by a federal act of expenditure or not. It does not mean that every increase in operations by such firms is related to governmental expenditures and the firms would not have made expansive expenditures without receipts from government. Resuming economic recovery and the expansion of firms are independently caused at this cyclical juncture.

Schumpeter claims it is inappropriate to attribute the resumption of prosperity and expansion to money inserted into the economy by government. What is "seen" is the income-generating expenditure and certain developments that are occurring simultaneously or subsequently. The relationship is not observed. Our interpretation of income-generating expenditure is either hypothetical or speculative, but not more so than an interpretation that relies solely on mechanical relationships between income-generating expenditure and the resumption of prosperity and expansion.

Interest rates on bonds (1932-1935), the wholesale price index (1926=100), and consumer price index (1967=100) for 1929-1940 are portrayed in Table 5.4.

From 1929 through 1933, there was a decline of 32% in the wholesale price index and a drop of 12.5% in the consumer price index. During the 1932 to 1935 period bond interest rates declined 24%. These periods of decline in the three quantities occurred despite the efforts of the Roosevelt administration to raise prices. The existence of underutilized resources and the pressure of increasing productive efficiency worked against a rise in prices. To the contrary, agricultural prices rose from an index of 40.9 in February 1933 to 78.3 in March 1935. The actions of the administration and Congress had a significant effect. Schumpeter warns against choosing dates that correspond with the decrease in the gold content of the dollar, as he thinks it would be meaningless.

Schumpeter observes that the demand for labor was reflected in an increase in the average wage rate of skilled and unskilled labor from 55 cents in 1933 to slightly over 64 cents in 1934. He attributes this to government policies that were enacted despite high prevailing unemployment.

FIGURE 5.4: U.S. NONAGRICULTURAL EMPLOYMENT AND FARM EMPLOYMENT

SOURCE: HISTORICAL STATISTICS OF THE UNITED STATES

Part I, pages 137, col. 127 and p. 468, col. 174

Figure 5.4 shows that farm employment declined starting in 1926. The decline was regular and rapid with increases in 1928-1929 and 1934-1935. This decline is indicative of the consolidation of farms, their increasing productivity, low farm prices for parts of the period, and sporadic weather conditions (such as poor crop seasons and the Dust Bowl). From 1916, farm employment demonstrated a downward trend. It was correlated negatively with personal income ($r = -.71$, $p = .000$, $N=41$). Interpreted, this implies that the increase in non-agricultural employment meant lower levels of personal income.

Schumpeter (1939) observes that while there was a 66% increase in output it was accompanied by a 33% increase in wages. He attributes this to the high levels of unemployment signifying underutilization of the work force. Non-farm employment declined 7.748 million persons from 1929 to 1932. Personal income and non-farm employment were highly correlated ($r = .94$, $p = 0.00$, $N = 41$). While causation cannot be inferred from this relationship, it is consistent with a demand theory hypothesis for the depression era.

Schumpeter's Disappointment with His Model

Schumpeter states that the difficulty of the model was not the fragility of the "prosperity" that occurred in the 1930-1932 period, but that it was followed by a deep depression that occurred at a rapid rate. At this point in his writing, Schumpeter questions whether the capitalist process had spent its force. He speculates that private investment opportunity had disappeared, making the economy dependent on government expenditure for motivating power. He feared that the economic system might collapse with the withdrawal of government expenditure.

Federal income-generating expenditures for 1935, 1936, and the first half of 1937 generated, directly or indirectly, affected prosperity in those years. The policy of increased expenditures finally took effect. During the first seven months of 1937 most of the $900 million of income-generating expenditures available was spent. A slump occurred in the latter part of 1937. A $4 billion deficit was budgeted for 1938. Schumpeter speculates that the slump would lead to recovery given the deficit spending. Schumpeter expected another decline would follow. Another decline followed in the first quarter of 1939. Schumpeter thought that there was little reason to expect that the mechanism of economic evolution would produce a recovery within the following several years.

Innovations: 1935-1939

New blast furnaces, coke ovens, open-hearth furnaces, electric furnaces, and new rolling mills all came into being and became operational. The prefabricated house (stimulated by the Federal Housing Act of 1936), domestic electrification, refrigerants, protective coatings, synthetic materials, and plastics all came into being. Rayon moved into new markets and found new uses. Air conditioning, although available since 1919, reached a total of about 17 million units in 1930. By 1937, there were approximately 85 million units in use. The airline industry was in the initial stages of development. Military demand, exports, and domestic demand all increased despite the presence of modest profits. Schumpeter argues that there was evidence of prosperity in 1937 and that prosperity would have exerted itself without impetus from government expenditure or other factors.

Exports and Restrictions

Exports in current dollars did not reach their 1929 levels ($5.241 billion) during the Great Depression era. Exports reached a low in 1934 ($1.280 billion) and proceeded to increase to $1.977 billion in 1937. General improvement in the world economy, devaluation, and demand for armaments accounted for most of the improvement.

Monetary Policy and Reserves

Changes in monetary policy and credit did not exert a major influence on the U.S. economy in the 1930-1937 period. While influence of money rates on price levels and credit occurred, the Banking Act of 1935 codified centralization of the Federal Reserve System. This development had the result of emphasizing restrictive rather than expansive monetary policy. Restrictions imposed on transactions in foreign exchange were removed in November 1934.

The increase in reserve requirements was a sign for member banks to initiate a reduction in their investments. Member reserve banks reduced their excess reserves from about $2.9 billion dollars on July 15, 1936 to $1.8 billion on August 19, 1936. Adjusted demand deposits decreased slightly in August and then proceeded to increase at a rapid rate. It is cause for questioning why the Federal Reserve Board did not leave the monetary situation as it was. A decline in federal income-generating expenditures was likely, and the influx of gold was managed by sterilization policy. A revival in business borrowing and a strong increase in money in circulation did not require further action of the Federal Reserve Board because of the phase the industrial process was in. Schumpeter maintains that mechanistic views about the money supply were regarded as more important than they really were. Announcement in late January 1937 of a 33.33% increase in reserve requirements could not, he maintains, be responsible for depressive symptoms. In particular, it was not responsible for the rapid decline of corporate security issues in the third and fourth quarters of 1937.

Capitalist Evolution and Sociological Circumstances

Schumpeter makes an effort to appraise the effects of a decline in income generation by government expenditure in 1937. The transition that occurred from prosperity to recession required a difficult reorientation of the economy with the attendant risk of breakdown. Severe slumps at such a transition time may be accompanied by excesses of speculation and, in particular, excess of speculation brought about by a fast pace of industrial development. These circumstances were obviously absent in the United States situation.

Schumpeter thought that income generation by government must always create issues of adaptation. The maximum income generation and its cessation by the United States government came at a time that was inappropriate. The collapse was brought about by a severe slowdown in income generation. The processes of recession remained sublimated. They failed to work in their traditional ways. This view implies the existence of additional and more fundamental issues.

This view is shared by many fellow economists. They offer as a theory the explanation that there was declining investment opportunity. This explanation occurs because of the occurrence of a decline after a weak prosperity (or "recovery"). In order to understand the Depression era, it is essential to reject this explanation.

The theory can be stated in terms of Schumpeter's analysis. Capitalism is in large part a process of endogenous economic evolution. Without economic evolution, capitalist society cannot exist. Capitalism would decline if innovations ceased to exist. The lack of entrepreneurs and entrepreneurial achievement and the lack of capitalist returns would mean that the elements propelling the capitalist system would be missing. Capitalism would be unable to survive under such conditions. The term "stable capitalism" is a contradiction in terms. Changes in production functions are essential to the capitalist process.

The proposition that the current (1938-1939) U.S. situation and capitalist process is stagnating, and income generation by the U.S. government is nothing more than a prop for a declining organism is inaccurate. Capitalism relies on the actual delivery of returns for its functioning and survival. The theory that investment opportunity is declining is not, by itself, adequate.

Schumpeter suggests that a more explanatory theory needs to be introduced. The working of capitalism produces a moral code that is hostile to it. Policies are produced that are hostile to the capitalist process. When the capitalist process is spending or has spent its momentum, there is no equilibrating process. What arises is a situation of possible deadlock in which neither capitalism nor possible alternatives work. This is what has happened in the United States Schumpeter emphasizes the presence of income taxes, corporate taxes, and estate taxes. He argues that the presence of those taxes on a group of from 30,000 to 40,000 high-income earners placed an important influence on "capital supply." He argues that these taxes exerted an important effect on business behavior. This effect, he maintained, was increased by failure of legislation permitting the carrying forward of business losses from one year to the next by new treatment of personal holding companies and other actions having an effect on actual or potential capital. Schumpeter observes that at issue is a matter of the value of assets, not of liquidity. It can be argued that accumulations by a business or industry make it easier for that firm to withstand the effects of a depression. Accumulations held in a liquid form may work in an anti-cyclic way. Sufficient book reserves are a necessary prerequisite for adequate amounts of raw material. Without these book reserves, businesses and industry would take a much more cautious policy towards expansion and continued production. Schumpeter observes a number of measures acted and interacted to combine to cause investment and capital formation to decline.

Schumpeter maintains that it is known that behind legislative measures, administrative actions, and expectations, there was something more fundamental. This something was a hostile attitude toward the industrial bourgeoisie, which was a product of the social process that created the bourgeoisie. The bourgeoisie feel threatened and are threatened. They are on trial before the judges and the public. They feel that their pocketbooks are threatened. An increasing part of the public is impenetrable to the viewpoints of business and industry.

Schumpeter maintains that by reflecting on major changes in the relationships between the individual and the state, as well as any shift in the favor of the state of revenue earned, basic changes in the habits of the mind, in attitudes, and the values of those immediately concerned are involved. He uses the English example to demonstrate how the bourgeoisie adapted to the principles of taxation over a period of thirty years. In the United States, Schumpeter maintains there was no such preparation. Because of the lack of preparation, there was a different response. Prior to the crisis the moral world of the businessman was the nation's moral world. The change in policy dates from 1934 to 1935. The radicalization of the public mind between 1930 and 1933 led, rather than followed, changing policies.

Schumpeter insists on the importance of personnel and methods of administration. New measures and new attitudes are best implemented by a skilled civil service. Roosevelt set out a task that was difficult for the most experienced bureaucracy. In the United States, a new bureaucracy was suddenly created. There was a minimum of experience and no clear idea of what a civil service is and what it could do. There was little experience; a lack of skill; and a lack of tact, reserve, and ability, which are expected from an experienced bureaucracy. Individuals and groups formed their own policies and advocated them with Congress and the public. Counsels of self-denial and patience were disregarded. Business was threatened by the aggressiveness and the lack of tact.

In concluding, Schumpeter maintains that there should not be much doubt about the ability of factors external to the process of economic evolution to account for many of the disappointing features of the 1937-1939 period. This includes the weakness of response to the system of government expenditure. The primary inadequacy of government expenditure was its lack of effect on unemployment and investment.

Schumpeter's prognosis was that there would be intermissions or reversals: 1) to the effects of "acclimatization"; (2) to the fact that (if his conceptual framework is to be trusted) recovery and prosperity phases would be more strongly delineated; and (3) recession and depression phases will be less strongly marked in the next three decades than they have been in the past two. He maintains, however, that the sociological context will not change.

The Author's Initial Views

Schumpeter's views on economic evolution, innovation, and equilibrium provide a creative way of addressing the issues of technological development and population change—although in the latter case, they are weakened by a lack of explicit definition. The study of macroeconomics tends to overlook the persistent process of coming into existence, growing, maintenance, and decline that is ongoing within the capitalist economic process. Schumpeter's theoretical concepts work in that direction.

His discussion of the concepts of equilibrium—development of the terms "near equilibrium" and "imperfect equilibrium"—appear to be a much more honest way of establishing the concept.

A disappointing aspect of this work is his effort to tie economic events and their development over their future horizons to a tripartite wave schema featuring Kondratieff, Jugular, and Kitchin waves. His work does not prove any connection of the economic forces he discusses to this wave schema. His exposition of this schema finally crumbles when he gets to the Depression and its later stages (1934-1938), although his work only briefly touches on 1938.

Part II will address this deficiency and more.

CHAPTER VI

WORLD WAR I AND GOLD AS REASONS FOR THE DEPRESSION

Introduction

Peter Temin, in his book, *Lessons from the Great Depression,* holds that the origins of the Depression were largely influenced by the disorder caused by World War I (WWI) and rigid adherence to the gold standard. He viewed the 30 years between 1916 and 1944 as a long conflict interrupted by an uneasy truce. The unemployment rates in the United States are indicative of this instability (Table 6.1).

Figure 6.1 graphically portrays the course of unemployment and the unemployment rate over the 1914-1940 period. From 1929-1930 and from 1930-1933, unemployment and the unemployment rate rose at a rapid rate. From 1929-1933, the average annual number of persons unemployed increased 2.82 million persons. This level of unemployment is adjusted in Part II of this book.

150 PHILIP S. SALISBURY

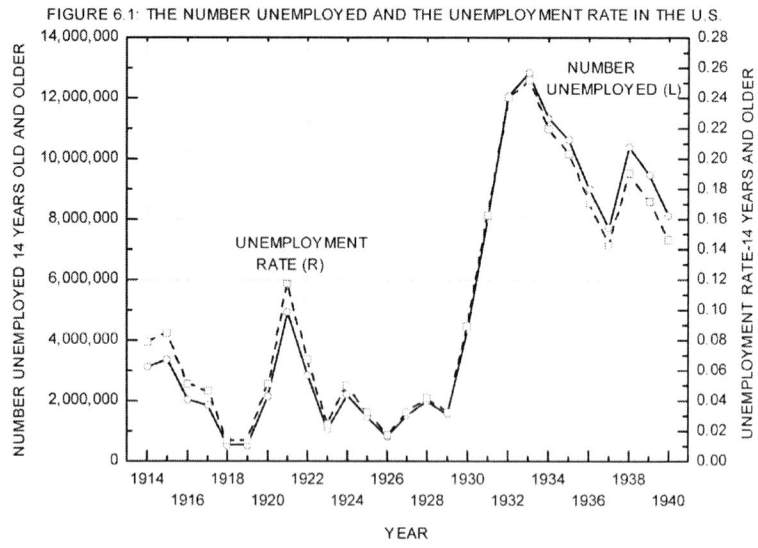

FIGURE 6.1: THE NUMBER UNEMPLOYED AND THE UNEMPLOYMENT RATE IN THE U.S.

SOURCE: ORIGINAL DATA FROM HISTORICAL STATISTICS OF THE UNITED STATES

Part 1, p. 126, col. 8, 9

Table 6.1
Unemployment and Unemployment Rate*
in the United States: 1916-1940

Year	Number Unemployed	Unemployment Rate
1916	2,043,000	5.1%
1917	1,848,000	4.6
1918	536,000	1.4
1919	546,000	1.4
1920	2,132,000	5.2
1921	4,918,000	11.7
1922	2,859,000	6.7
1923	1,049,000	2.4
1924	2,190,000	5.0
1925	1,453,000	3.2
1926	801,000	1.8
1927	1,519,000	3.3
1928	1,982,000	4.2
1929	1,550,000	3.2
1930	4,340,000	8.9

1931	8,020,000	16.3
1932	12,060,000	24.1
1933	12,830,000	25.2
1934	11,340,000	22.0
1935	10,610,000	20.3
1936	9,030,000	17.0
1937	7,700,000	14.3
1938	10,390,000	19.1
1939	9,480,000	17.2
1940	8,120,000	14.6

Unemployment and Unemployment Rate* in the United States: 1916-1940, *unadjusted, from Historical Statistics of the United States, Part 1, p. 126, col. 8, 9

Gold Reserves

WWI strained the gold standard. WWI led to the suspension of the gold standard, but not to its termination. After WWI, there was a move in the United States as well as in England, France, and Germany to revive the gold standard of the pre-WWI period. The unstable interwar gold standard was affected by WWI. The Depression was not an inevitable consequence of WWI. The shock of WWI, and efforts to reestablish the gold standard were important factors generating the Depression. Temin maintains that the Depression was not unavoidable in 1929. If policymakers had been able to free themselves from their dedication to the gold standard they could have enacted countercyclical policies.

The presence of the gold standard dictated inflation. Deflationary forces were accented in 1929. Holding an industrial economy to the gold standard further exacerbated deflationary tendencies.

Table 6.2
United States Gold Stock in Billions, Annual Change in Billions, Percent Change, and Personal Income in Billions

Year	Gold Stock* In Billions	Annual Change In Billions	Annual Percent Change	Personal Income in Billions
1916	$2.556	$.5307	20.76%	$33.7
1917	2.868	.3122	10.89	62.5
1918	2.873	.0490	.17	62.5
1919	2.707	-.1658	-6.12	65.0
1920	2.639	-.0684	-2.59	73.4
1921	3.373	.7346	21.78	62.1
1922	3.642	.2685	7.37	62.0
1923	3.957	.3151	7.96	71.5
1924	4.212	.2556	6.07	73.2
1925	4.112	.1101	-2.43	75.0
1926	4.205	.0926	2.20	79.5
1927	4.092	-.1128	-2.76	79.6
1928	3.854	-.2379	-6.17	79.8
1929	3.997	.1425	3.57	85.9
1930	4.036	.3096	7.19	77.0
1931	4.173	-.1334	-3.20	65.9
1932	4.226	.0529	1.25	50.2
1933	4.036	-.1904	-4.72	47.0
1934	8.285	4.2225	51.13	54.0
1935	10.125	1.8672	18.44	60.4
1936	11.423	1.2965	11.35	68.6
1937	12.790	1.3675	10.69	74.1
1938	14.592	1.8015	12.35	68.3
1939	17.800	3.2080	18.02	72.8
1940	22.042	4.2422	19.25	78.3

* Source: Historical Statistics of the United States, Gold Stock, Part 2, page 995, col. 438; **Personal Income, Part 1, p. 224, col. 8

England abandoned the gold standard on September 20, 1931. The shift in policy allowed the British economy to lower export prices relative to foreign prices. It also permitted the British to establish a more relaxed monetary policy. The Bank of England did not allow gold to flow out of their country. Thus, England built up its reserves. British actions had a deflationary impact on the rest of the world. Britain uncoupled its future from countries that adhered to the gold standard and permitted its policy to be less limiting. An inflow of gold from the United States into Britain occurred in 1931.

An internal drain on gold in 1931 increased the number of Federal Reserve notes outstanding. Law specified that the Federal Reserve hold 40% gold and an additional sum of collateral in the amount of 60% in either gold or eligible paper.

Needed in the Depression were economic forces and policies that were expansionary in nature. Adhering to the gold standard was not compatible with that objective. The United States lived by the gold standard until March 6, 1933. Roosevelt proclaimed a bank holiday on that day. The proclamation also prohibited banks from dispersing gold or dealing in foreign exchange. The Emergency Banking Act was passed on March 9, 1933. This act extended the March 6th official announcement and gave the President emergency powers covering banking transactions, foreign exchange dealings, and gold and currency transactions. On March 10, 1933, an executive order was issued extending the restrictions on gold and foreign exchange dealings after the bank holiday. The order prohibited gold payments by both banking and non-banking institutions. An exception was made for transactions licensed by the Secretary of the Treasury. An executive order was issued on April 5, 1933, which prohibited "hoarding" of gold and required holders of gold coin, bullion, or certificates to deliver these to Federal Reserve Banks by May 1st. There were exceptions made for reasonable amounts for industrial use, rare coins, and a maximum of $100 dollars per individual in gold coin and gold certificates. Gold coin and certificates were exchanged at face value. Bullion was purchased at the market price of $20.67 per fine ounce. December 28, 1933, marked the completion of the "nationalization" of gold. An order was issued by the Secretary of the Treasury at that time. Again, the reimbursement price for gold bullion, gold coin, and gold certificates was $20.67. Later, the expiration date for the surrender of gold was reset as January 17, 1934. At that time the market price of United States gold was set at $33 a fine ounce.

Table 6.2 and Figure 6.2 show the annual changes in the amount of gold stock and the amount of gold stock held by the U.S. Treasury. The re-pricing of gold by the United States, gold inflows into the United States because of higher prices, and the nationalization of the gold supply resulted in a 51.13% increase in U.S. gold stocks held at the end of 1934.

Why was the nationalization of gold undertaken? President Roosevelt clearly stated that his administration would encourage the dollar to depreciate against foreign currencies as a method of causing domestic prices to rise. An amendment was offered to the Agricultural Act was enacted into law on May 12, 1933. The amendment was explicitly directed at causing a price rise through the expansion of the money stock. It contained a provision authorizing the President to reduce the gold content of the dollar by as much as 50% of its former weight. The dollar price of gold started rising. Similarly, the dollar price of foreign currencies began to rise. In the period between May 12, 1933, and January 29, 1934, the market price of gold rose and then fluctuated between $27 and $35. The Gold Reserve Act was passed on January 30, 1934. In that interim period, the United States had a floating exchange rate. Starting September 8, 1933, an official gold price was fixed daily at an estimated world market figure with adjustments for shipping and insurance.

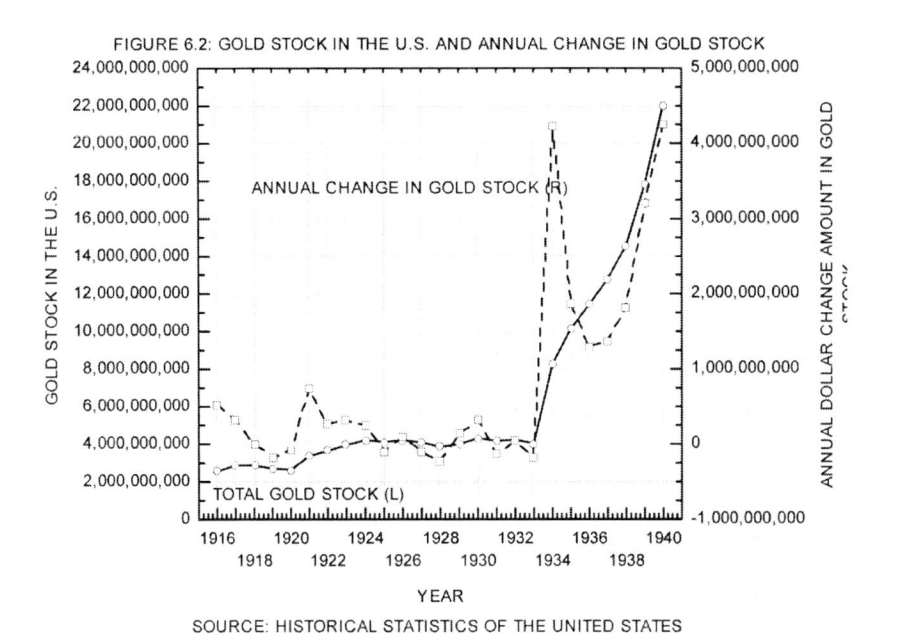

FIGURE 6.2: GOLD STOCK IN THE U.S. AND ANNUAL CHANGE IN GOLD STOCK

SOURCE: HISTORICAL STATISTICS OF THE UNITED STATES

Part 2, page 995, col. 438

In October, 1933, the U.S. government actively intervened to raise the gold price. The Federal Reserve Banks started purchasing gold internationally shortly after the RFC was authorized to purchase newly mined domestic gold. From November 1933 to the end of January 1934, the price offered exceeded the overseas market price.

Gold policy was focused on raising the price level of commodities. Commodity prices had endured the largest relative decline in the preceding deflationary period. The Agricultural Adjustment Administration's use of production controls was an example of another approach to this issue. Commodity price increases were not pursued through substantial increases in the quantity of money, although a legal basis for that existed. The decline in the foreign exchange value of the dollar featured a roughly proportionate increase in the price of domestic farm commodities. The purpose of gold policy to increase prices of farm products and raw commodities was, for the most part, achieved. The decline in the exchange value of the dollar encouraged exports and discouraged imports. These forces were reinforced by the purchase of domestic and foreign gold by the United States.

Gold represented one class of commodities. Other countries, especially those on a gold standard, were impacted by the U.S. revised gold price. Those countries were committed to sell gold at a fixed price with their own currency pegged at the level of exchange. These countries necessarily encountered pressure on their gold reserves. This, in turn, made it necessary for these countries to abandon the gold standard or have to endure internal inflationary pressure. Aside from the changes in the relative supply and demand for goods that were imported or exported as a result of the gold purchase program, countries found it necessary to adjust their nominal price level for goods.

Friedman (1963, p. 468) draws attention to the existence of a major obstacle to using gold to lower the exchange value of the dollar. A gold clause was present in many government and private obligations in contracts. This clause required repayment in either gold or a nominal amount of currency equivalent to the value of a specified weight of gold. On June 5, 1933, Congress passed a resolution that abolished the legislation establishing the gold clause. The action was later upheld by the Supreme Court.

The period of a variable price for gold was terminated on January 31, 1934, when, under the Gold Reserve Act, the president specified a fixed buying and selling price of $35 an ounce for gold. The gold dollar was devalued to 59.06% of its former weight. Title to all gold coin and bullion was to be held permanently by the United States. Gold coins were to be withdrawn from circulation and melted into bullion. Any additional gold coinage was to be discontinued. The revaluation of the dollar caused the United States Treasury to realize a profit of approximately $3 billion without the purchase of any additional gold. The weight of gold required to be held declined. Gold certificates could be legally held by Federal Reserve Banks, but not by individuals. In order to achieve its paper profit, the Department of the Treasury printed approximately $3 billion of gold certificates without purchasing any additional gold and turned the gold certificates over to the Federal Reserve System. The Department of the Treasury received, in return, a deposit credit that it could convert into Federal

Reserve notes or draw on by check. The economic consequence of this transaction was to provide the Treasury with the authority to print approximately $3 billion of fiat currency. This supplemented $3 billion in currency authorized by the Thomas amendment to the AAA. Two billion dollars of the paper profit was placed in a stabilization fund for use of the Secretary of the Treasury for purposes of stabilizing the exchange value of the dollar.

From February 1, 1934, through December 31, 1940 (and some time thereafter), the price of gold remained fixed at $35 an ounce. Over that time period the state of gold reserves did not have an effect on policy. In the late thirties most of the gold-bloc countries abandoned the gold standard. As an alternative, they adopted nominally floating exchange rates with government use of stabilization funds.

In summary, Temin argues that the Great Depression was caused by the excessive stresses of WWI and the 1919-1932 period on the gold standard. The war led to the suspension of the gold standard, but not to its extinction. The system of gold standards was revived in its prewar form, although the reestablishment of the gold standard did not come easily. The asymmetry of the gold standard between countries was hard on those nations lacking reserves. Nations lacking reserves found it necessary to contract while nations rich in gold reserves expanded. Gold standard policies encouraged deflation for the reserve-rich nations and devaluation for the reserve poor nations. The consequence in the late 1920s was world deflationary policy. Gold standard instability was the consequence of WWI. The Depression was not a certain consequence of the war. The unyielding policy institution of the gold standard generated the Great Depression.

Author's Comment

The following figure (Figure 6.3) allows some insight into the relationship of gold stock and the consumer price index (1914=100).

FIGURE 6.3: GOLDSTOCK COMPARED TO THE CONSUMER PRICE INDEX: 1914=100

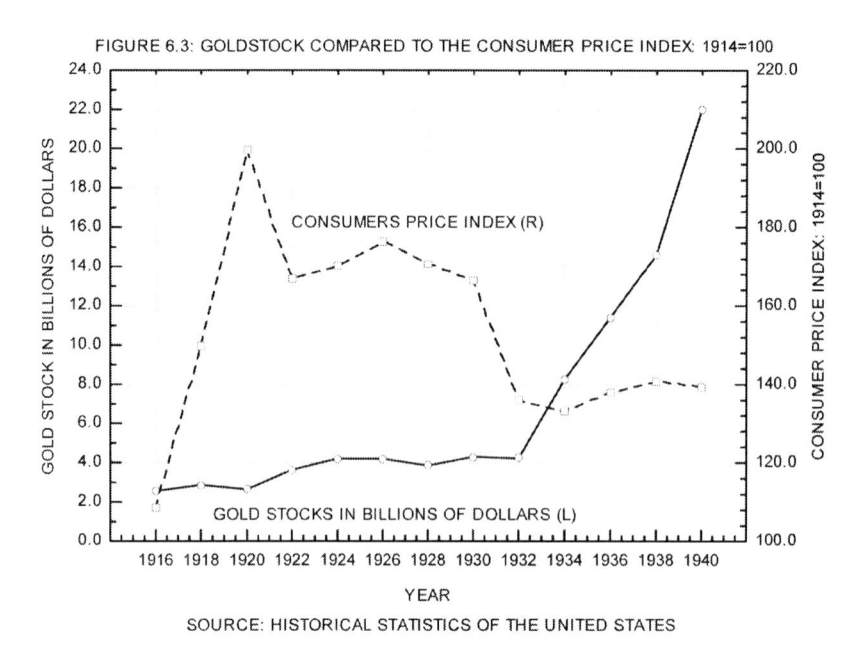

SOURCE: HISTORICAL STATISTICS OF THE UNITED STATES

CPI, Part1, p. 164, col.727, Gold stock, Part 1, p. 200, col. 40

What is of importance about this view of gold stock and the Consumer Price Index (CPI) is that the gold stocks were not significantly statistically related to the CPI in any major way from 1916-1940. One may conjecture, however, that it is true the low level of gold stocks from 1916-1932 placed a deflationary drag on the economy during that time. There is clearly no statistical relationship between these two variables but the graphic evidence supports this view. Furthermore, gold stock is uncorrelated to personal income, and output. There is no significant relationship between M_2 supply and gold stock. The insignificant statistics do not resolve the gold hypothesis as gold *may* have had a stabilizing impact on inflation after 1934.

Before rejecting the gold hypothesis, however, another comparison needs to be made. The comparison is between gold stock and personal income. Table 6.2 and Figure 6.4 allow that comparison.

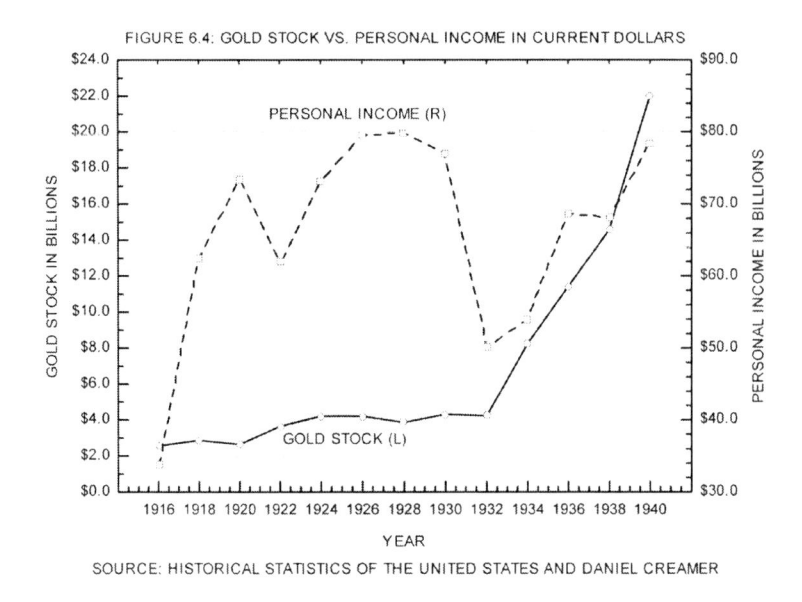

FIGURE 6.4: GOLD STOCK VS. PERSONAL INCOME IN CURRENT DOLLARS

SOURCE: HISTORICAL STATISTICS OF THE UNITED STATES AND DANIEL CREAMER

Personal Income, Part 1, p. 224, col. 8; Gold stock Part 1, p. 200, col. 40

While gold stocks increased from 1932 through 1940, so did personal income. A different situation is portrayed from 1916 through 1930. Personal income remains high during that time but appears to be unaffected by gold stocks. Does this represent, as Temin suggests, the discontinuance of the gold standard during WWI and the gradual efforts to reestablish the gold standard, plus the rigid adherence to the gold standard up to February 1934? The deflationary impact of going off the gold standard and revaluing the gold to $35 an ounce must be admitted to have had possible effects on personal income from late 1933 forward.

Eichengreen and Sachs

Eichengreen and Sachs (1985, 1986) view the interwar standard as the principal carrier of deflation. They conclude devaluation or abandonment of the gold standard was a first step in national and world recovery. In their research (1985), they examined differences between countries abandoning the gold standard at an early time (at or before

1931) and those countries abandoning the gold standard later. They concluded that countries leaving the gold standard early had much more rapid recoveries than those remaining on the gold standard. Consistent with their perspective was that the gold standard led to a monetary contraction.

Eichengreen and Sachs (1985) tested an hypothesis about the role of wages in determining aggregate supply. They used a sample of ten industrial countries in 1935 with the output variable industrial production. They found a strong negative relationship between output and real wages across the ten countries. Countries that fell into the high real wage, low output category were countries that remained on gold well after 1931.

Bernanke

Bernanke (2000) had a larger sample (N=22, 1931-1936) than did Eichengreen and Sachs (1985, 1986) to test the view that leaving the gold standard gave nations a greater ability to increase money supply. The view presented was that staying on the gold standard promoted the deflationary process. Bernanke (1995, 2000) advanced the view that an essential first step out of the Depression to national and global economic recovery was going off the gold standard.

Countries included in the high-real-wage, low-output area are all countries that remained on gold beyond 1931. These nations included Belgium, France, Italy, and the Netherlands. Nations with low real wages and higher output included Denmark, Finland, Norway, Sweden, and United Kingdom. Bernanke could not find any real theory to explain this phenomenon.

Bernanke also found evidence of nominal wage stickiness in the Depression and indicated sticky wages were a dominate source of non-neutrality.

Bernanke's view (2000) based on more extensive data while reinforcing of Eichengreen and Sachs (1985, 1986) leaves Bernanke to state:

> . . . we are not aware of any plausible story of why these declines in spending should have affected so many disparate countries around the world so nearly simultaneously and in particular why they should have been more persistent in countries remaining on the gold standard. (p. 286)

As a contrast, this paper suggests that total personal income in the U.S. is reduced in response to the declining magnitude of selected population subgroups (54-75+ years) in the 2007-2021+ period.

CHAPTER VII

KEYNES AND EFFECTIVE DEMAND

Definitions

Keynes in his book *The General Theory of Employment, Interest, and Money* (1936), provides an early definition of effective demand. He maintains:

> the volume of employment is given by the point of intersection between the aggregate demand function and the aggregate supply function; for it is at this point that the entrepreneur's expectations of profits will be maximized . . . . the point of the aggregate demand function, where it is intersected by the aggregate supply function, will be called *effective demand* (p. 25).

Alternatively stated, effective demand is the aggregate proceeds the entrepreneurs expect to receive, including incomes they will transfer on to other factors of production as well as employment which they decide upon. The aggregate demand function relates hypothesized quantities of employment to the amounts that accrue as an expected consequence of profits yielded by their outputs. Effective demand is a legitimate concept because taken in combination with conditions of supply, it corresponds to the level of employment that maximizes the entrepreneur's profit expectations. Put in the symbolic terms of a stress Equation:

$AS_t = AD_t$ at or during equilibrium	(7.1)
AS_t = aggregate supply at or during time t	
AD_t = aggregate demand at or during time t	

Written as a stress Equation:

$S_t = AS - AD_t$ (7.2)
S_t = stress at or during time t; it may be either positive, negative, or zero (equilibrium) depending on the relative values of AS_t and AD_t

It needs to be pointed out that this Equation is neither a difference Equation nor a deviation Equation in its truest form although it has roots in the latter. It is not a difference Equation in that it does not subtract successive terms from each other. It is the close cousin to the deviation Equation but does not use the mean as the subtrahend. The equilibrium or steady-state values are inserted in place of the mean. In place of the subtrahend are the equilibrium or steady-state values for the total economy or a subset of goods and/or services.

Net income is the amount we expect the person of average income to recognize as the available income he or she has when making his consumption choices (now disposable personal income). Alternatively, saving is the excess of income over the costs of consumption. This necessitates a definition of consumption. Consumption is defined as the value of goods and services sold to consumer-purchasers or investment-purchasers at or during a defined time period.

The amount of saving is an outcome behavior of individual consumers and the amount of collective investment of individual entrepreneurs. Provided income is equivalent to the current value of output, that current investment is equivalent to the non-consumed part of current output, and saving agrees with its previous definition. The equality of saving and investment follow. In word Equations (Keynes, 1936, p. 63):

Income = value of output = consumption + investment
Saving = income - consumption
Therefore,
Saving = investment

the equality between the amount of saving and the amount of investment comes from the dual character of the transactions between the producers of goods and services and the consumer or purchaser of goods and services.

Author's Additions

Following up on Keynes' definitions of aggregate supply and aggregate demand, the definition of their equality at equilibrium is turned into a Stress Equation. In turn, the Stress Equation can be changed into a partial equilibrium model or a general equilibrium model. Additional steps will be taken with a more detailed expression of aggregate demand:

$$AD_t = \sum_{z=1}^{z=m} (\overline{Npy_{zt}} - \overline{Ntx_{zt}} + \overline{Nsv_{zt}} + \overline{Ndb_{zt}}) + \sum_{g=1}^{g=r} \overline{GOV_{gt}} + \sum_{d=1}^{d=f} \overline{GDPI_{dt}} + \sum_{q=1}^{q=e} \overline{NEX_{qt}} \quad (7.3)$$

Made dynamic:

$$dAD = d\sum_{z=1}^{z=m} (\overline{Npy_z} - \overline{Ntx_z} + \overline{Nsv_z} + \overline{Ndb_{zt}}) + d\sum_{g=1}^{g=r} \overline{GOV_g} + d\sum_{d=1}^{d=f} \overline{GDPI_d} + d\sum_{q=1}^{q=e} \overline{NEX_q} \quad (7.4)$$

$$\overline{dt} \quad \overline{dt} \qquad \qquad dt \qquad dt \qquad dt$$

The meanings of the terms in Equation 7.3 and Equation 7.4 are as follows:

AD_{xt} = total aggregate demand for goods and services (x) at any time t or during any time period t = 1 . . . n

N = the number of individuals in group z at or during time t

py_{zt} = the average personal income for group z; z = 1 . . . m, at or during time t

tx_{zt} = the average taxes of group z at or during time t

sv_{zt} = the average savings of group z acquired (-) or withdrawn (+) at or during time t

db_{zt} = the average debt acquired (+) or paid off (-) during time t

In Equations 7.3 and 7.4, expenditures on taxes and savings are treated as part of personal income and are taken out of personal income (the first term of the Equation),

$\sum_{g=1}^{g=r} GOV_{gt}$ = governmental expenditures of governmental units 1 through r at or during time t

$\sum_{d=1}^{d=f} GDPI_{dt}$ = gross domestic private investment of firms and organizations 1 through f at or during time t

$\sum_{q=1}^{q=e} NEX_{qt}$ = net exports by the nations q=1 to q=r at or during time t

Equations 7.3 and 7.4 set the stage for discussion and conclusions that will be presented in Part II, Chapter IX, X, XI, and XII, and Appendix II. Also important to this discussion is a definition of aggregate supply at or during time t (AS_t). Aggregate supply of goods and services is the entrepreneur's (businesses, industries, and not-for-profits) anticipation of the aggregate level of demand he or she can expect, including the demand produced by marketing efforts.

In economic theory and practice, aggregate demand and aggregate supply (output) move towards equilibrium as time passes. There are levels of aggregate output, which exceed aggregate demand, that are equivalent to aggregate demand, or are insufficient to meet aggregate demand. Substituting in the stress equation basic to Stress Theory, one could assume one of two alternatives.

$$Eq_{xt} = AD_{xt} \text{ and } A_{xt} = AS_{xt} \qquad (7.6)$$

Alternatively,

$$Eq_{xt} = AS_{xt} \text{ and } A_{xt} = AD_{xt} \qquad (7.7)$$

What Stress Theory demonstrates is that *there is no dependent variable other than S_{xt}.* The level of stress is dependent on the relationship of AD_t and AS_t. This makes it mathematically possible for both variables (AD_t and AS_t) to be influenced by *variation in their components*. In addition, each of the variables, *aggregate supply and aggregate demand,* **may** *vary independently of each other. The dependent variable in this Equation is the stress variable* (S_{xt}). It allows the possibility of independent variation of AD and AS that causes disequilibria (stress or asymmetry) in an economy. Both aggregate demand and aggregate supply may be subject to a large number of variations that create disequilibria. For purposes of this discussion, the terms AD_{xt} and AS_{xt} will be substituted in the stress Equation as follows: $AD_{xt} = Eq_{xt}$ and $AS_{xt} = A_{xt}$.

$D_t = \sum_{x=1}^{x=m} d_{xt}$ Units of demand (D_t) equals the sum of the demand for goods and services at or during time t (d_{xt}) including each category of goods and services x at or during time t

OP_t = general output price level of goods and services at or during time t

DP_t = general demand price of consumers and purchasers of goods and services at or during time t

Output (quantity) and output prices for aggregate supply are related to units (quantity) of demand and demand price from all sources in the following way in equilibrium:

$$\sum_{x=1}^{x=m} o_{xt} \, opr_{xt} = O_t Pr_t = AS_t = AD_t = DPr_t D_t = \sum_{x=1}^{x=m} dpr_{xt} d_{xt} \qquad (7.8)$$

It is noteworthy that the central equation from Equation 7.8 may not be balanced. In that instance, $AS_t \neq AD_t$. What is important about Equations 7.3 and 7.4 is that each of the originating terms of the equation has been disaggregated into one interpretation of its components. The lack of disaggregation in the use of these terms can result in economists missing the currents underneath the surface of the terms aggregate supply and aggregate demand.

It is important to recognize that in a steady-state an economy's stress may vary in a range about the equilibrium value of $S_{xt} = 0$. In the steady-state an economy is varying about equilibrium while being neither in recession nor in a period of pronounced growth. As will be demonstrated later, cyclical variations may be influenced by variation in disposable personal income. Variation in personal income and personal consumption is influenced by the variations in total disposable income by specific, mutually exclusive subgroups.

Aggregate supply is the total value of goods and services that businesses and industries are willing to produce in a defined time period. Aggregate supply is a function of inputs, technology used, and price. Aggregate demand is a function of the following Equation (7.9):

Aggregate Demand =	Personal Consumption +	Government Capital Expenditures +	Gross Domestic Private Investment +	Net Exports +

More precisely, the terms of Equation (7.9) are disaggregated in Equations 7.10 and 7.11.

In turn, the stress Equation can be changed into a method for disaggregating each of the terms of Equation. Additional steps will be taken to provide a more detailed expression of aggregate demand that demonstrates the impact of the working age population on aggregate demand:

$$AD_t = \sum_{z=1}^{z=m}(\overline{Npy}_{zt} - \overline{Ntx}_{zt} + \overline{Nsv}_{zt} + \overline{Ndb}_{zt}) + \sum_{g=1}^{g=r}GOV_{gt} + \sum_{d=1}^{d=f}GDPI_{dt} + \sum_{q=1}^{q=e}NEX_{qt} \qquad (7.10)$$

Made dynamic,

$$\frac{dAD}{dt} = \frac{d\sum_{z=1}^{z=m}(\overline{Npy}_z - \overline{Ntx}_z \pm \overline{Nsv}_z \pm \overline{Ndb}_{zt})}{dt} + \frac{d\sum_{g=1}^{g=r}GOV_g}{dt} + \frac{d\sum_{d=1}^{d=f}GDPI_d}{dt} + \frac{d\sum_{q=1}^{q=e}NEX_q}{dt} \qquad (7.11)$$

The meanings of the terms in Equation 7.10 and Equation 7.11 are as follows:

AD_t = total aggregate demand for goods and services at any time t or during any time period t = 1 ... n

N = the number of individuals in group z at or during time t

py_{zt} = the average personal income for individuals in group z; z = 1 ... m, at or during time t

tx_{zt} = the average taxes per individual in group z at or during time t

sv_{zt} = the average savings per individual in group z at or during time t (+) or withdrawal from savings (−)

db_{zt} = the average debt acquired per person at or during time t (+) or average debt paid off (−) at time t

$\sum_{g=1}^{g=r} GOV_{gt}$ = governmental capital expenditures of governmental units g; g = 1 ... r, at or during time t

$\sum_{d=1}^{d=f} GDPI_{dt}$ = gross domestic private investment of firms and organizations d; d = 1 ... f, at or during time t

$\sum_{q=1}^{q=e} NEX_{qt}$ = the dollar value of net exports by the nations q; q = 1 ... e at or during time t

In Equations 7.10 and 7.11, expenditures on taxes, savings, and debt are treated as part of personal income and are taken out of personal income (the first term of the Equation),

$$\sum_{z=1}^{z=m} (\overline{Npy}_{zt} - \overline{Ntx}_{zt} - \overline{Nsv}_{zt} + \overline{Ndb}_{zt}) \qquad (7.12)$$

Also important to this discussion is a definition of aggregate supply at or during time t (AS_t). Aggregate supply of goods and services is the entrepreneur's (businesses, industries, and not-for-profits) production of goods and services in anticipation of the aggregate level of demand (an expected level of demand), including the demand produced by marketing efforts, and the secondary demand for goods and services.

CHAPTER VIII

MONEY SUPPLY AND BANKING

Friedman and Schwartz

Friedman and Schwartz's view that greater liquidity was needed during the Depression bears some scrutiny. To test the role of money supply, the relationship of personal income (PINC), the Wholesale Price Index (WPI), the adjusted unemployment rate (ADJUNR), and M_2 were tested. The following statistics resulted (N = 41).

Table 8.1: Correlation Matrix

	ADUNR	M_2	WPI
M_2	.42*	—	
WPI	NS	.49*	—
PINC	NS	.95**	.68**

* $p \geq .001$, **$p = .000$

Later Equations (10.2 and 10.3, Chapter X) suggest that the relationships of M_2, personal income, the adjusted unemployment rate, and wholesale price index are intertwined in the estimation of the size of the money supply and secondly in the determination of personal income. These equations, it will be repeated, are, the author believes, indicative of an equilibrating balance between the two variables with modifications by the adjusted unemployment rate and the wholesale price index. Without further discussion of that perspective, the status. of banking, deposits, currency, and gold stocks will be addressed.

Money, Banking, Deposits, Currency, and Gold Stocks

The current definitions of money supply are reflected in the values of M_1 and M_2. M_1 is a narrowly defined set of the money supply and consists of coins, paper currency, plus all demand or checking deposits. This is money supply used in transactions. M_2 is a broadly defined money supply. It includes M_1 plus liquid assets or near-monies such as saving deposits, money market funds, and certificates of deposits.

A relationship exists between a price rise and a fixed nominal supply. This condition produces tight money and results in lowering of aggregate spending. This is a condition known as the money-supply effect. It is maintained by some economists to be importantly responsible for the Great Depression.

The major proponents of the view that money supply and its regulation were a primary cause of the Great Contraction are Friedman and Schwartz (1963, p. 300).

> The monetary collapse was not the inescapable consequence of other forces, but rather a largely independent factor that exerted a powerful influence on the course of events.
>
> The failure of the Federal Reserve reflected not the importance of monetary policy, but rather the particular policies followed by the monetary authorities, and, in smaller degree, the particular monetary arrangements in existence.
>
> The contraction is in fact a tragic testimonial to the importance of monetary forces. . . . . different and feasible actions by the monetary authorities could have prevented the decline in the stock of money—indeed, could have produced almost any desired increase in the money stock.
>
> But it is hardly conceivable that *money income* could have declined by over one-half and prices by over one-third in the course of four years if there had been no decline in the stock of money. (Emphasis mine)

Let's take a look at some of the graphic and empirical evidence (see Figure 8.1 and Table 8.2). Table 8.2 indicates there was a period of increasingly tight money from 1929 to 1933. Money supply levels per person in the civilian labor force were less in those years than in 1928. Recovery in the 1934-1940 period did not achieve 1928 levels until 1940. Care must be taken in interpreting this because of possible errors in estimating the civilian labor force.

An important aspect of M_2 is its relationship to the adult civilian labor force. Figure 8.1 and Table 8.1 reveal a period of somewhat tight money in 1929 through 1934. A trend towards a loosening of the money supply started in 1934. Whether this period of five years was adequate to initiate the depths of the Depression is open to question. Figure 8.1 confirms the decline in M_2 from 1930 to 1933 and its increase from 1934 to 1940. It should be noted that all three quantities (M_2, the currency in circulation, and the gold stock) are on the increase from 1934 to 1940. The period from 1929 through 1933 is a partial result of individuals reducing their deposits in banks, hoarding, unemployment, discouraged workers, being on the gold standard, the collapse of banks, and possibly the tightness of the money supply.

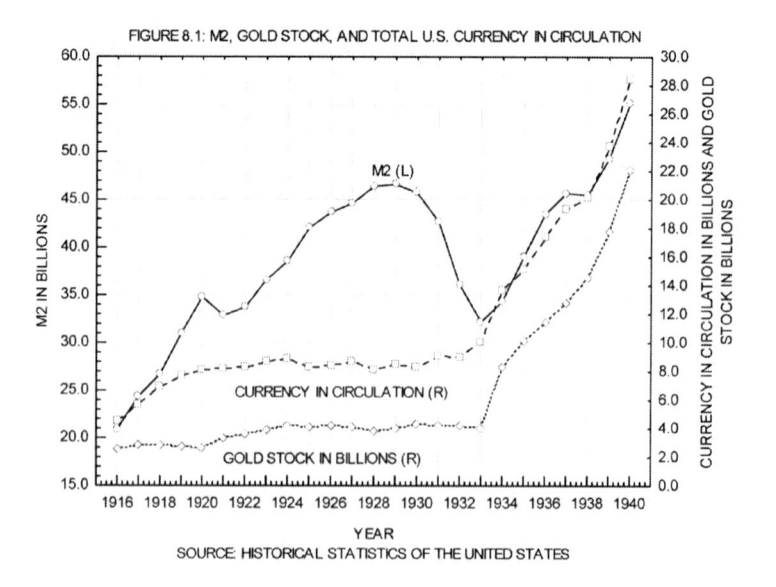

FIGURE 8.1: M2, GOLD STOCK, AND TOTAL U.S. CURRENCY IN CIRCULATION

SOURCE: HISTORICAL STATISTICS OF THE UNITED STATES

M_2, Part 2, p. 992, col. 415; CIC, Part 2, p. 993, col. 423; Gold Stock, Part 2, p.995, p. 438

Table 8.2 and Figure 8.1 suggest that the growing magnitude of M_2 is inadequate to accommodate the growing civilian labor force. This supports analysis that suggests that money supply tightness and decline are major contributors to the Great Depression. The marked increase in these three money measures—currency in circulation, gold, and M_2 after 1933 (see Figure 8.2) is a complement to the slow moving recovery of the 1934 through 1940 period.

However, the high level of adjusted unemployment (unemployed and discouraged/total labor force) (29.0% in 1932 and 31.1% in 1933) suggests that there was little demand for money. The level of deposits per bank did not start increasing until 1934 after the FDIC law was enacted. The impact of this entity on bank suspensions started being noticed in 1934.

Table 8.2: Civilian Labor Force and M_2/Person in CLF

Year	Civilian Labor Force (Millions)	M_2/CLF Person	Year	Civilian Labor Force (Millions)	M_2/CLF Person
1920	41.000	$848.78	1931	50.440	$846.35
1921	42.034	781.51	1932	51.102	705.46
1922	42.671	790.22	1933	54.526	699.65
1923	43.708	837.37	1934	52.226	657.51
1924	43.800	880.82	1935	52.786	740.16
1925	45.406	926.08	1936	53.432	813.75
1926	44.500	981.57	1937	53.846	848.34
1927	46.030	971.75	1938	54.684	832.23
1928	47.190	983.67	1939	55.116	893.93
1929	48.437	962.06	1940	55.163	992.51
1930	49.885	916.71			

Sources: Calculations and data based on Historical Statistics of the United States CLF, Part 1, p. 126, col. 4; M_2, Part 2, p. 992, col. 415

Banking

The following data (Figure 8.2 and Table 8.3) on the number of banks and the reduction in their numbers through failure and consolidation reveal the lack of confidence depositors came to have in those institutions. Not only were banks under great stress, the number of depositors was greatly reduced due to several of the factors mentioned previously. One factor new to this mix of reasons is the decline in the number of foreign-born males, 15-74 years of age, from 1930 to 1940 (see Table 13.3). The rushes on banks can be seen as initiated or reinforced by this loss of a significant pool of depositors. A reduction in the number of depositors represented a major reason for contraction in the banking industry.

In 1920, there were 30,909 banks: 22,885, or 74% of these, were non-national banks. The decline to 15,076 total banks in 1940 (65.7% non-national banks) represented a decline of over 51% of all banks from 1920. Fifty-seven percent of the non-national banks and 36% of national banks were lost in the Great Depression era. Smaller banks were at much greater risk of defaulting or going out of business.

U.S. banks, which were Federal Reserve members and non-Federal Reserve member banks, offer insight into the impact of the FDIC and its introduction in 1934. Table 8.3 records bank suspensions for both classes of banks.

Table 8.3: National and Non-National Banks 1929-1940

Year	All Banks	Non-national Banks Non-national	Percent
1929	26,401	18,038	68.3%
1930	25,568	25,568	66.6
1931	24,273	15,442	63.6
1932	22,242	13,172	59.2
1933	19,317	9,874	51.1
1934	15,913	10,496	66.0
1935	16,047	10,622	66.2
1936	15,884	10,516	66.2
1937	15,646	10,353	66.2
1938	15,419	10,177	66.0
1939	15,210	10,007	65.8
1940	15,076	9,912	65.7

Source: Historical Statistics of the United States
All Banks, Part 2, p. 1019, col. 580; Non-national banks, Part 2, p. 1026, col. 656

Banking had an irregular path both prior to the Great Depression and during the Great Depression. Banks were one of the institutional focal points of the Great Depression. All banks increased in number from 13,053 in 1900 to a peak of 31,076 in 1919. From 1919 on, all banks decreased in number to 15,076 in 1940. Non-national banks increased from 9,322 in 1900 to a high of 22,926 in 1921. That was followed by a decrease to 9,874 non-national banks in 1933 (see Figure 8.2).

The decline from 1924 to 1933 was approximately 52%. The banking industry was subjected to a contraction in the number of banks in early 1933. A bank holiday in the United States followed from March 6, 1933, to March 13, 1933. This holiday was designed to stem the rush on banks by depositors eager to preserve their balances.

The contraction in the number of all banks extended from 1924 to 1940 and beyond. Liquidations, mergers, consolidations, as well as failures contributed to the decline in all banks from 30,444 in 1924 to 15,076 banks in 1940. In addition to the decline in banks from 1933 to 1940, the decline in the number of all banks continued after 1940. Not only did bank failures result, but a process of consolidation occurred.

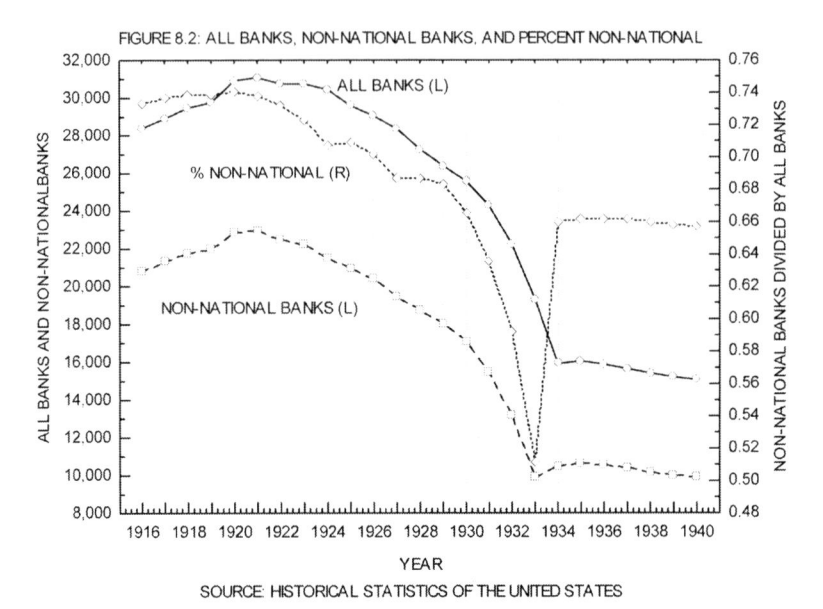

FIGURE 8.2: ALL BANKS, NON-NATIONAL BANKS, AND PERCENT NON-NATIONAL

SOURCE: HISTORICAL STATISTICS OF THE UNITED STATES

All banks, Part 2, p. 1019, col. 580; Non-national banks, Part 2, p. 1028, col. 656

The differentiation between U.S. banks and non-national banks reveals a significant difference between the number of banks in each category that were suspended in the 1930 through 1940 period. In the case of the smaller non-national banks, their reserves were less than U.S. banks, making them more susceptible to failure or suspension than U.S. banks. In the case of non-national banks, some of the foreign reserves they held were subject to inflation and soon became worthless (see Table 8.4 for data on bank suspensions).

Table 8.4: Suspended United States Banks
and Suspended Non-national Banks, 1929-1937*

Year	Number of United States Banks Suspended	Number of Non-national Banks Suspended
1929	64	547
1930	161	1,104
1931	409	1,697
1932	276	1,085
1933	1,101	2,616
1934	1	43
1935	4	30
1936	1	42
1937	4	52

* Source: Federal Reserve Bank of St. Louis

The process of federal auditing and certifying banks for reopening that was started in 1933 was reinforced by the introduction of the FDIC. Its ability to insure the deposits of depositors up to $100,000 went a long way towards bringing back the confidence of potential depositors. The remaining issue was that of individuals having enough money to deposit. Deposits grew from $41.68 billion in 1933 to $70.85 billion in 1940.

M_2 and the Unemployment Rate

The correlation of M_2 and the adjusted unemployment rate is weak to moderate (r = .42, p = .006, N = 41). This is illustrated by Figure 8.3.

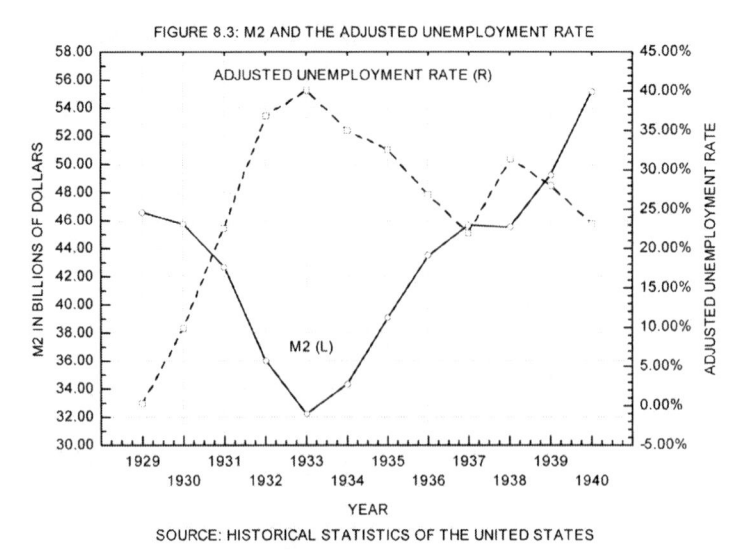

FIGURE 8.3: M2 AND THE ADJUSTED UNEMPLOYMENT RATE

SOURCE: HISTORICAL STATISTICS OF THE UNITED STATES

CLF Part 1, p. 126, col. 4; Calculations based on p. 126 col. 8; M₂ Part 2, p. 992, col. 415

This is insufficient evidence (N=11) to justify an inverse univariate relationship between M_2 and the adjusted unemployment rate.

GNP, Velocity, and M_2

The relationship between Gross National Product (GNP) (in current dollars) and M_2 is a strong one (r=.97, p=.000, N=51) (see Figure 8.4). During the period from 1931-1933, M_2 decreased at a smaller percentage rate than did GNP. From 1934-1940, M_2 rose at a rate exceeding the rate of recovery of GNP. This is an indication that increasing the supply of M_2 during the 1930-1939 period may have had a beneficial effect on GNP. A third often-neglected concept is that of velocity. Fisher (1906, 1911) developed the equation:

$$MV = PQ \qquad (8.3)$$
M = money supply
V = velocity
*P = price
*Q = quantity
*(PQ = GNP)

Figure 6.3 demonstrates that velocity declined approximately 3.4 times from 1890 to a low of 1.61 in 1932. Velocity increased to 1.98 times in 1937, and then declined to 1.81 times in 1940.

The declining trend line of velocity from 1890 through 1932 had an important impact on GNP (in addition to increasing productivity). This, in combination with M_2, would be enough to cause a significant trough in GNP. It should be noted that the trough in the velocity of money in 1932 preceded the trough in GNP 1933 by one year.

From 1900 to 1940, velocity and the unemployment rate were associated with each other ($r = -.81$, $N = 41$, $p < .000$). The nature of the relationship of M_2, velocity, and their relationship to GNP is displayed in Figure 8.4.

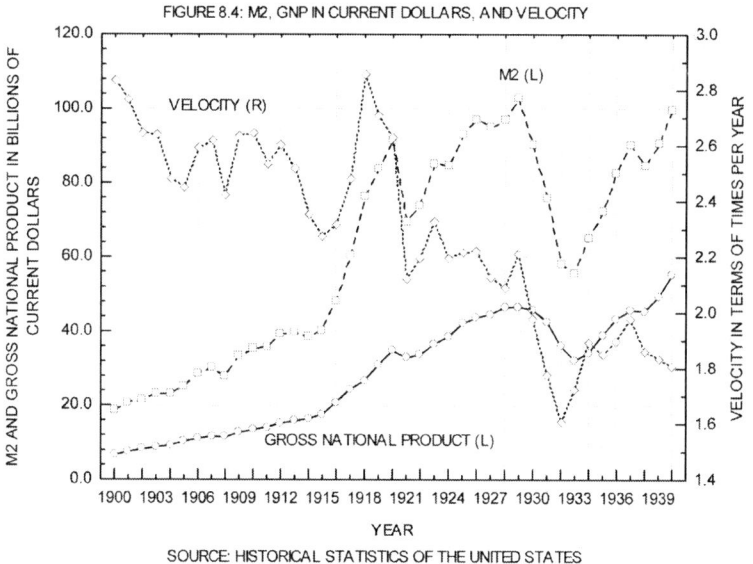

FIGURE 8.4: M2, GNP IN CURRENT DOLLARS, AND VELOCITY

SOURCE: HISTORICAL STATISTICS OF THE UNITED STATES

GNP, Part 1, p. 224; M_2, Part 2, p. 992, col. 415, Velocity calculated

Table 8.5: Velocity of Money, M_2, GNP in Current Dollars, and Adjusted Unemployment Rate, 1929-1940

Year	Annual Velocity (in #times/year)	M_2 (in billions)	GNP (in billions)	Adjusted* Unemployment Rate
1929	2.21	$46.60	$103.1	5.0%
1930	1.98	45.73	90.4	10.4
1931	1.78	42.69	75.8	19.1
1932	1.61	36.05	58.0	29.0
1933	1.73	32.22	55.6	31.1
1934	1.89	34.36	65.1	27.6
1935	1.85	39.07	72.2	25.9
1936	1.90	43.48	82.5	22.0
1937	1.98	45.68	90.4	18.6
1938	1.86	45.51	84.7	24.9
1939	1.84	49.27	90.5	22.6
1940	1.81	55.20	99.7	19.3

M_2, Part 2, p. 992, col. 415; GNP, Part 1, p. 224, col. 1; *calculated from the number unemployed, and the number of discouraged workers, and the adjusted civilian labor force

The following figure (Figure 8.5) demonstrates the strength of the relationships between bank deposits and bank reserves. Bank deposits are correlated with bank reserves at $r = .84$ ($p = .000$, $N = 41$).

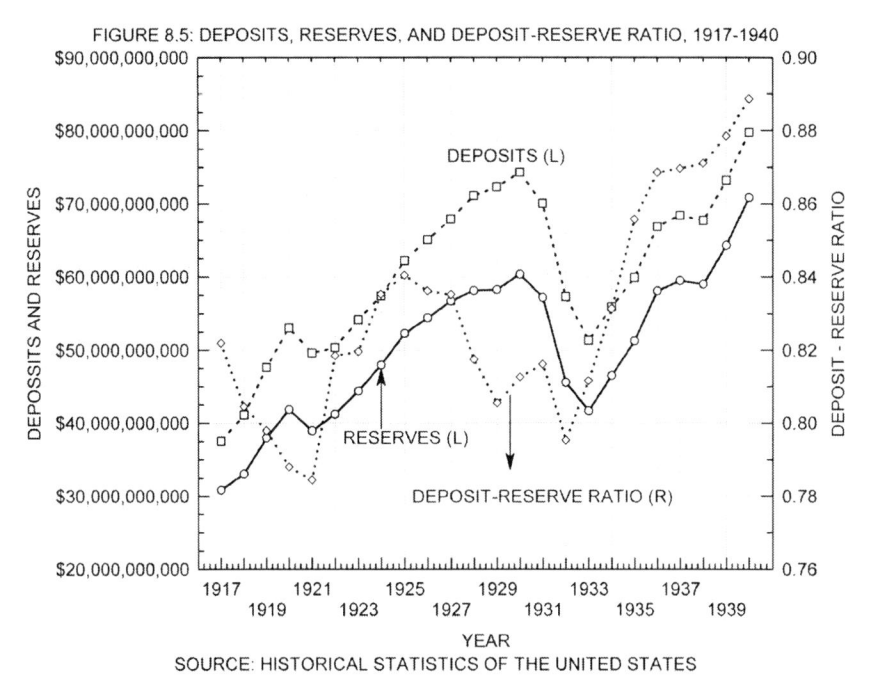

FIGURE 8.5: DEPOSITS, RESERVES, AND DEPOSIT-RESERVE RATIO, 1917-1940

SOURCE: HISTORICAL STATISTICS OF THE UNITED STATES

Deposits, Part 2, p. col. 796; Reserves, Part 2, p. 1042, col. 802

Deposits

Deposits dropped at a rapid rate starting in 1931. Deposits returned to their 1930 levels in 1939. Deposits declined from $60,365 million in 1930 to $41.684 million in 1933. This was a decline of approximately 31% (see Figure 8.5).

Deposits per bank peaked in 1930 at approximately $2.36 million. The trough in deposits was approximately $2.05 million in 1932. The drop in deposits per bank (13.2%) was a harbinger of the banking holiday in 1933. The level of deposits per bank rose to $4.7 million per bank in 1940 as the number of all banks declined.

In the period from 1900 to 1921, the deposit-reserve ratio ranged from .78 to .82. From 1929 to 1933, the deposit-reserve ratio ranged from .795 to .816, with a low of .795 in 1932. The Banking Act of 1933 provided temporary, but extensive, protection for depositors. In August of 1935, a permanent deposit insurance system became effective due to the provisions of Title I of the Banking Act of 1935. From 1935 to 1940, the deposit-reserve ratio rose from .856 to .889. In addition, the deposits per bank rose from $3.2 million to $4.7 million in 1940 (an increase of $1.5 million per bank). This evidence is a testimonial to the success of the Banking Act of 1935.

Observing Figure 8.5 and associated data, deposits and reserves reached a Depression era low of $41.684 and $51.359 billion, respectively, in 1933. The deposit-reserve ratio, however, reached a minimum of .795 in 1932. A recovery of sorts began for banks in 1934 after the implementation of the FDIC program.

Currency and Loans

Total currency rose from $2.37 billion in 1900 to $28.5 billion in 1940. Total currency continued to increase from 1900 to 1927. From 1927 to 1930 the decrease was approximately $3.61 million. From 1931 to 1932, there was a lesser decline of $75.1 thousand. These periods of decline coincided with the decline in the number of all banks. This compares to an increase of $18.38 billion from 1933 to 1940. Reflected in these data are greater confidence in the banking system, an increase in money supply, and the increased acquisition of gold.

Another significant indicator of banking activity is that of loans by banks. From the peak in 1929 ($41.94 million) to the trough in 1935 ($20.24 million), there was a 52% drop in the face value of loans made. This is a definite reflection of the poor health of the economy, the struggling banking system, the tightening of credit, and declining personal income. Some of the correlations of loans with other variables in the period from 1900 to 1940 follow with their significance values (N=41).

Table 8.6: Correlations with Loans (N=41)

Variable	Loans (in Current Dollars)
Output	.88*
M_2	.87*
Currency in Circulation	.65**
TotalCurrency	.41(p=.008)
CPI(1967=100)	.92*

*p=.000, **p=.001

Of note are the periods after 1919, in which the annual percentage of M_2 was declining. These periods in the U.S. economy were times of slowing economic activity or economic hardship. The period of 1920-1921 endured economic recession. M_2 decreased $1.95 billion, or 6.0%. From 1929-1933 the value of M_2 declined. From 1929 through 1933 M_2 declined $14.38 billion, or approximately 31%. M_2 increased from a low of $32.22 billion in 1933 to a high of $45.68 billion in 1937. A period of recession followed in 1938, as M_2 declined to $45.51 billion dollars. In 1939 and 1940 M_2 increased. Periods of declining M_2 are ominous to those concerned with the health of the U.S. economy.

The declining purchasing power of money during the Depression era is evidenced by the changing consumer price index (1914 = 100) (see Table 8.7).

Table 8.7: Consumer Price Index, 1920-1940 (1914 = 100)

Year	CPI	Year	CPI
1920	199.7	1931	151.5
1921	178.1	1932	136.1
1922	166.9	1933	128.8
1923	169.7	1934	133.3
1924	170.3	1935	136.7
1925	174.8	1936	138.1

1926	176.2	1937	143.1
1927	172.8	1938	140.6
1928	170.9	1939	138.4
1929	170.9	1940	139.5
1930	166.4		

Source: Historical Statistics of the United States, Part 1, p. 164, col. 727

The consumer price index declined 27% from 1926 to 1934. This is a 27% drop in the value of a market basket of goods. Likewise, there is a 27% decline in money supply over the same period. The Consumer Price Index from 1934 on gradually increased, but remained below its 1929 levels. M_2 and the CPI from 1926 to 1934 reinforce those who view money supply as generating some of the dislocation of the Great Depression. Relevant correlations of the CPI with M_2 and current dollar GNP follow:

Table 8.8: Correlation Matrix
CPI, GNP, and M_2 (N=51)

	CPI	M_2
M_2	.88*	—
GNP	.94*	.97*

* $p \leq .000$

CHAPTER IX

FARMING

Farming Trends

Employment on farms declined from a peak of 13.555 million persons in 1907 to a low of 10.979 million persons in 1940. Non-farm employment increased at the expense of farm employment. Immigration from foreign countries as well as migration from the south to the north contributed to non-farm employment. Immigration contributed a much lesser extent to farm employment. From 1910 to 1940, non-farm employment increased from 21.7 million persons in 1907 to a high of 37.98 million persons in 1940. Farm employment fluctuated around the 10.979 to 12.818 (with family workers) million-person level from 1930 to 1940 (see Figure 9.1 and Table 9.1).

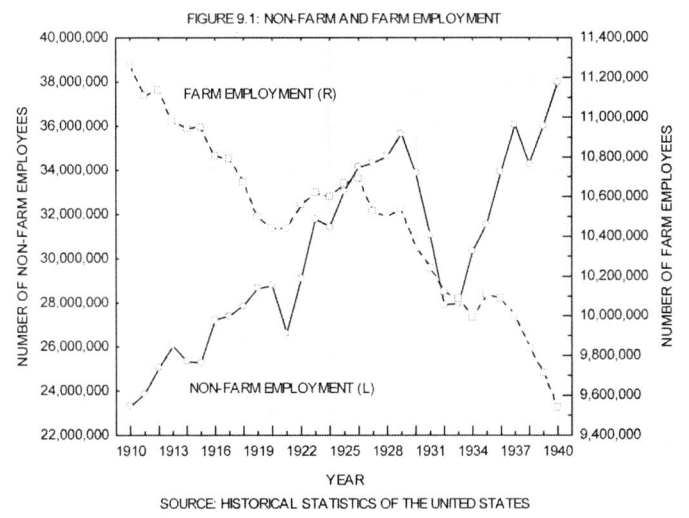

Farm employment, Part 1, p 126, col. 6; Non-farm employment, p. 126, col. 7

Non-farm employment witnessed periods of brief decline in 1908, 1921, and in the 1929-1932 period. The 1929 non-farm employment level of 35.67 million persons was reached in 1937 with a recession in 1938 and a recovery in 1939-1940. In 1900, the farm employment level was 69% of non-farm employment levels. In 1940, farm employment was 25% of non-farm employment. This represented a decrease of 1.51 million employees on farms and an increase of 22.7 million non-farm employees from 1900 to 1940.

Table 9.1: Non-farm and Farm Employment in Millions

Year	Non-farm Employment	Farm Employment
1929	35.67	10.54
1930	33.84	10.34
1931	31.07	10.24
1932	27.92	10.12
1933	27.96	10.09
1934	30.32	9.99
1935	31.56	10.11
1936	33.90	10.09
1937	36.07	10.00
1938	34.30	9.84
1939	36.03	9.71
1940	37.98	9.54

Source: Historical Statistics of the United States,
Farm employment, Part 1, p 126, col. 6;
Non-farm employment, p. 126, col. 7

In relative terms, the increase in non-farm employment was more dramatic.

The increase in farms from 1880 to 1920 was from 4.01 million farms to 6.45 million farms. This was an increase of 2.44 million farms. From 1920 to 1940, farms decreased from 6.45 million to 6.1 million. The decrease in the number of farms from 1929 on was due to forfeiture as well as consolidation. The process of consolidation and increasing farm productivity is in evidence when one examines the change in the number of farm acres per farm employee (see Figure 9.2).

THE CURRENT ECONOMIC CRISIS AND THE GREAT DEPRESSION   187

FIGURE 9.2: FARM ACREAGE PER EMPLOYEE AND INCOME PER FARM

SOURCE: HISTORICAL STATISTICS OF THE UNITED STATES

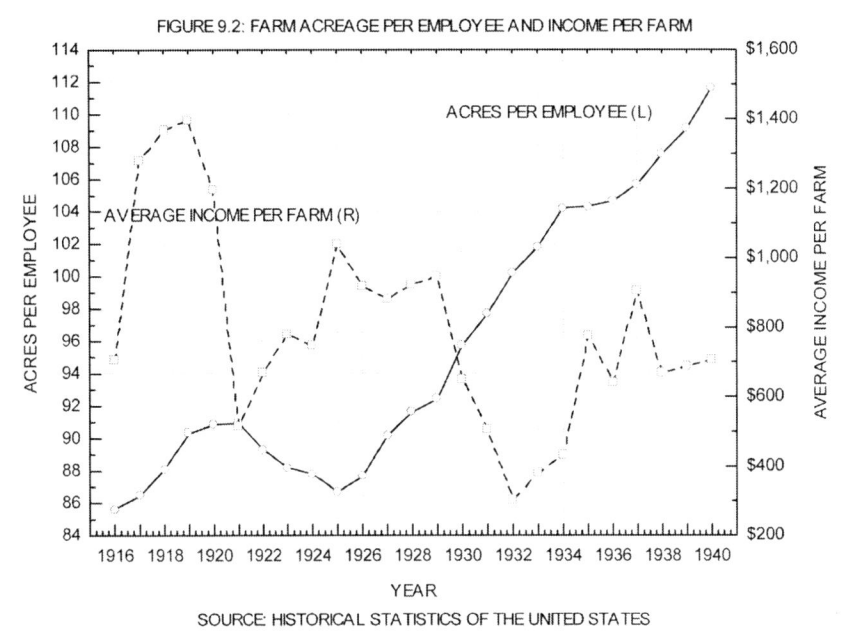

Land in farms, Part 1, p. 457, col. 5; Farm employment, Part 1, p 126, col. 6; Average income per farm, Part 1, p. 483, col. 260.

The increase in farm acres per employee (see Figure 9.2) was a consequence of increased productivity, consolidation, foreclosures, and increasing farm acreage. Figure 9.3 shows declining farm employment with increasing farm acreage. Farming was the most productive sector of the U.S. economy in the early 1900s through 1924.

In 1929, GDP per work-hour for the entire economy began to decrease (see Figure 8.3). from 1929 through 1933, gross domestic product per man-hour for all employees declined by approximately $6.50 per hour. Farm GDP per hour per employee dropped precipitously. GDP per work-hour for farm employees remained at low levels until 1940. This is a consequence of the low levels of income per farm, increased productivity per employee (see Figure 9.2), and a variation in crop conditions. Figure 8.4 details the number and type of farms and farm ownership. Of particular note is the drop in the number of farms from 1920 to 1930 (-159,000) and from 1930 to 1940 (-193,000). Full owners of farms decreased from 1920 to 1930 (-455,094), and increased from 1930 to 1940 (+172,439). The decline in the number of farms from 1920 to 1930 was indicative of some of the problems farms were having in the 1920s, particularly low prices.

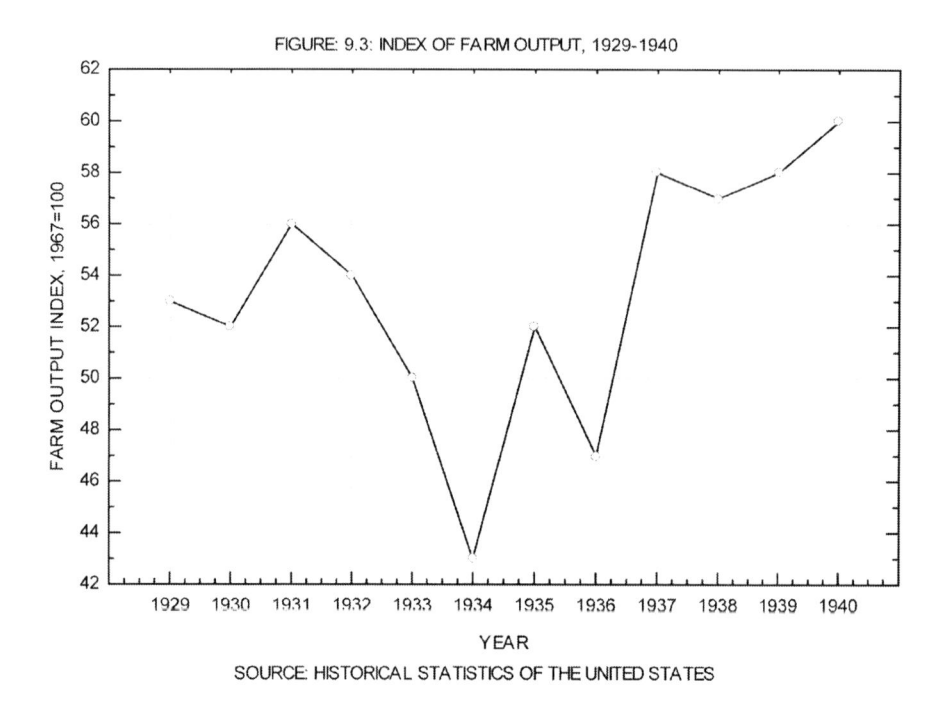

FIGURE: 9.3: INDEX OF FARM OUTPUT, 1929-1940

SOURCE: HISTORICAL STATISTICS OF THE UNITED STATES

Examining Table 9.2, there are several changes of note. The number of farms increased from 1900 to 1920. From 1920 to 1940, farms declined in number by 352,000. The larger part of that decrease (193,000 farms) came in the 1930 to 1940 period. From 1920 to 1930 full-owner farms declined by 455,094, or 14%. Full-owner farms increased slightly (+172,439). Two other categories will be described. Tenants increased from 1920 to 1930 (+210,257, or 9%) and declined from 1930 to 1940 (-303,888, or -11%). The remaining category, African-American tenants, decreased from 1920 to 1930 (-32,884, or -3%), and declined from 1930 to 1940 by 192,096, or -21%.

Table 9.2: Farms and Types of Farm Ownership*

YEAR	FARMS	FULL-OWNER	PART-OWNER	MANAGER	TENANT
1900	5,740,000	3,202,643	451,515	59,213	2,026,286
1910	6362,000	3,355,731	593,954	58,353	2,357,784
1920	6,454,000	3,368,146	558,708	68,583	2,458.554
1930	6,295,000	2,913,052	657,109	56,131	2,668,811
1940	6,102,000	3,085,491	615,502	36,501	2,364,923

* Source: Historical Statistics of the United States: P. 465, Series K 110-113; 123.

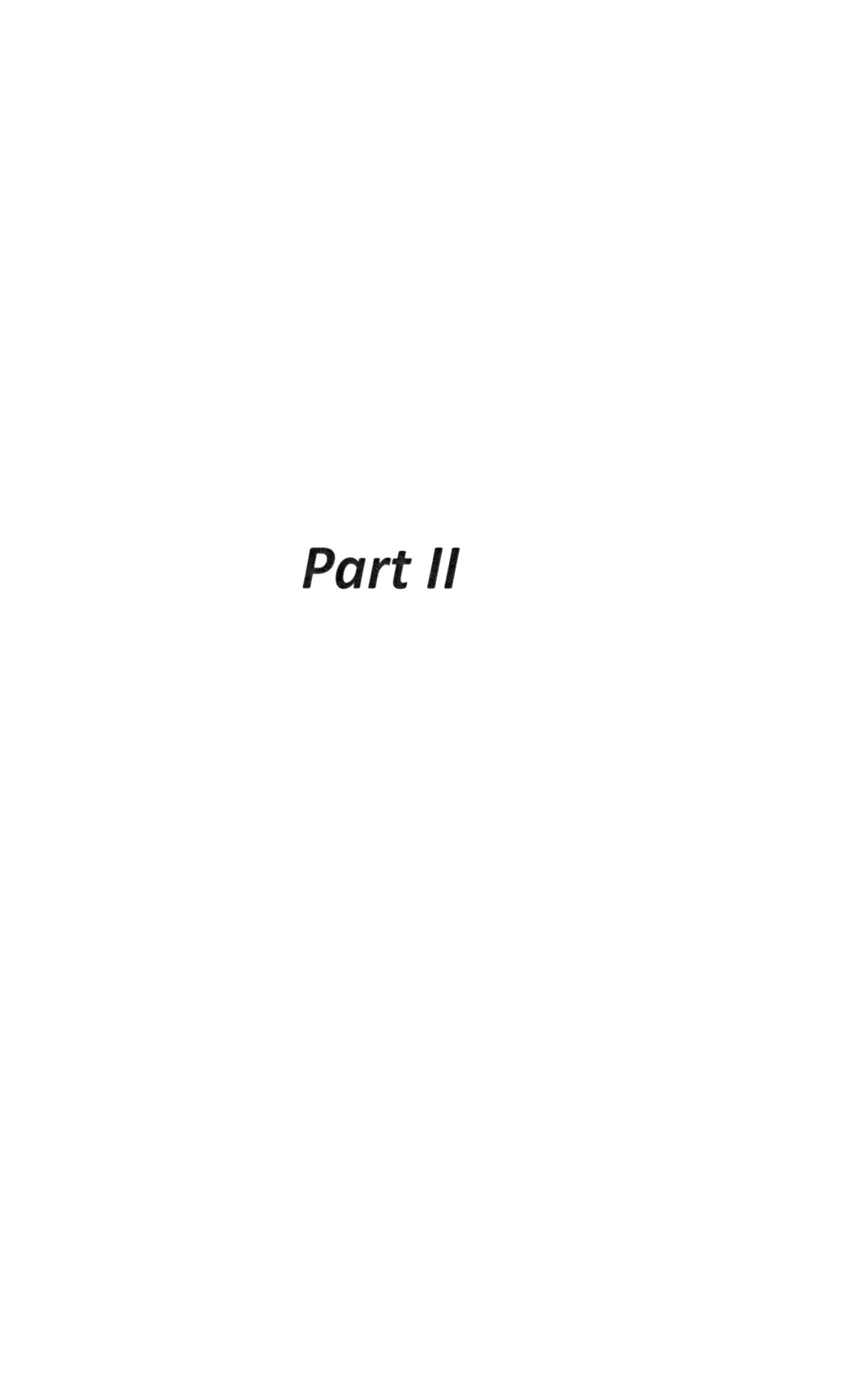

Part II

CHAPTER X

THESES AND SOME TESTS

A Major Feature of Economic Life

One of the major, but unheralded, features of economic life is the presence of macro-values that can be further disaggregated into subgroups or sub-quantities. One of the reasons this is not done is that this process adds further complexity to analyses and requires that adequate data be collected in a proper form. The primitiveness of data collection in the era of the Great Depression inhibits a proper diagnosis of population data and certain economic data during that era. This is a major drawback to understanding the importance of sex, employment rates, age, and average income levels to national quantities of personal income, disposable personal income, personal saving, and personal consumption. Also inhibited is an understanding of the impact of discouraged individuals on employment and unemployment.

Labor Force Estimates

Labor Force estimates contained in *Historical Statistics of the United States, Colonial Times to 1970* were based on those of Stanley Lebergott (1964). Employment and unemployment numbers for each age-sex-nativity group were interpolated between decennial rates from 1900 to 1930. Lebergott adjusted the 1910 count for over-count. Interpolation between 1930 and 1940 decennial totals were adjusted using the Bureau of Labor Statistics (BLS) total labor force series. The BLS series was developed by applying annual worker rates for age-sex groupings to census data for the appropriate groups.

Preliminary annual labor force and employment estimates were derived by interpolating between detailed worker rates in census years, and applying the resultant series to unpublished census estimates of population annually from 1900 to 1930.

The worker rates were interpolations between estimated 1930 labor force rates and those shown for 1940 by the Current Population Survey. The resultant series reflects changing proportions among the various age-sex groups . . . . Because of the widespread use of BLS figures and because the differences are well within the error involved in the computation of the duplicating item, Lebergott adopted the BLS figures beginning in 1930 as his unemployment totals, and then subtracted these from the labor force totals to get employment (Historical Statistics of the United States, author's analysis).

This methodology for developing labor force estimates, particularly in the 1930 through 1940 period, leaves much to be desired. In fact, it calls into question the estimates for that period. In spite of this deficiency of data and the ability to conduct statistical tests, a general theory will be offered that proposes to account for variations in aggregate demand during the Great Depression era. The mathematical logic of this argument supplemented by available statistical evidence will be relied on to present the author's views.

Theses

The central thesis (stated in positive form) of this theoretical and empirically based argument is: the number of individuals of given subgroups (most probably sex, age, educational level, and foreign-born or native-born status during the Great Depression era) influences aggregate demand and supply as well as disaggregated demand and supply through seven mechanisms:

1. The variation of the number of individuals (of income earning age) by specified subgroups that differ significantly on average personal income affects total real personal income for the subgroup, and therefore personal consumption, in a developed economy. This variation has its source in the number of individuals

born in each subgroup, their aging process, supplemented by the net immigration into and out of each subgroup.

2. There is variation in the *rates* of individuals in each subgroup who have personal income. This rate, the personal income participation rate (PIPR), or employment rate for each subgroup, affects personal consumption and, therefore, aggregate demand.

3. Differences in real *personal income by subgroup* affect total real personal income for each subgroup and, therefore, total personal income for the population.

4. Changes in subgroup real personal income may act as a leading influence on economic growth and decline as measured by changes in personal consumption and aggregate demand. Also affected are changes in levels of savings (and debt), aggregate demand, aggregate supply, changing output, and changing employment and unemployment.

5. Wage and salary levels of individuals are "sticky." Thus, periods of economic decline are generally not associated with major downward shifts in wages. Long periods of economic decline, such as depressions, periods in which there are large, relative influxes of immigrants or nonworking poor into the labor market, may put more significant downward pressures on domestic wages.

6. Stress Theory provides a theoretical and applied approach for evaluating equilibria, disequilibria, and "stress" for different components of the economy, such as aggregate supply and aggregate demand. Stress Theory is particularly useful for empirically evaluating what Fisher (1911) termed "transition" periods or periods of asymmetry.

7. The terms for personal income or personal consumption—government capital expenditure, gross domestic private investment, and net exports—also can be disaggregated. Their disaggregated variation also affects aggregate demand and aggregate supply.

$$\sum_{x=1}^{x=m} (\overline{Npy}_{xt} - \overline{Ntx}_{xt} - \overline{Nsv}_{xt}) = \text{average personal income of each subgroup x minus average taxes per person in subgroup x minus average savings per person in subgroup x (personal consumption)}$$

$$\sum_{g=1}^{g=r} GOV_{gt} = \text{governmental capital expenditures of governmental units 1 through r at or during time t}$$

$$\sum_{d=1}^{d=f} GDPI_{dt} = \text{gross domestic private investment of business firms and organizations 1 through f at or during time t}$$

$$\sum_{q=1}^{q=e} NEX_{qt} = \text{net exports by the nations q=1 to q=r at or during time t}$$

Estimating Discouraged Workers

The statistics available on the Depression era do not have an estimate of discouraged workers. To remedy this situation, an estimate was made of discouraged workers using monthly Current Population Survey data from January 1996 through December 2005 (N = 120). The data were not seasonally adjusted.

The results of this process yielded numbers of discouraged workers from 1929 through 1940. This was accomplished by adding discouraged workers to the number unemployed and calculating a revised unemployment level and revised unemployment rate. The unemployment rate and the number of discouraged workers were correlated at $r = .71$, $p = 0.00$, (N = 120). The following equation resulted:

$$\sum_{x=1}^{x=m} (\overline{Npy}_{xt} - \overline{Ntx}_{xt} - \overline{Nsv}_{xt}) = \text{average personal income of each subgroup x minus average taxes per person in subgroup x minus average savings per person in subgroup x (personal consumption)}$$

$$\sum_{g=1}^{g=r} GOV_{gt} = \text{governmental capital expenditures of governmental units 1 through r at or during time t}$$

$$\sum_{d=1}^{d=f} GDPI_{dt} = \text{gross domestic private investment of business firms and organizations 1 through f at or during time t}$$

$$\sum_{q=1}^{q=e} NEX_{qt} = \text{net exports by the nations q=1 to q=r at or during time t}$$

Table 10.1: Correlation Coefficient (N=120)

	Unemployment Rate
Discouraged Workers	.71*

* p = 0.00

Equation 10.1 and Statistics

Discouraged Workers = (75,644.6)(Unemployment Rate) - 9,691.6 (10.1)
(t = 10.88)   (p = 0.00)

F = 118.3, p = .000, SEE = +55,494
N=120

Variable	Mean	Standard Deviation
Unemployment Rate	4.9	.73
Discouraged Workers	363,867	78,469

Table 10.2: Unemployed Adjustment for Discouraged Workers, 1930-1940

Year	Discouraged Workers	Unemployed	Total (1+2)	Revised Unemployment Rate
1930	666,890	4,430,000	5,006,890	10.4%
1931	1,220,252	8,020,000	9,240,252	19.1
1932	1,811,287	12,060,000	13,871,287	29.0
1933	1,897,702	12,830,000	14,727,702	31.1
1934	1,651,121	11,340,000	12,991,121	27.6
1935	1,525,395	10,640,000	12,135,395	25.9
1936	1,278,659	9,.030,000	10,308,659	22.0
1937	1,073,598	7,700,000	8,773,598	18.6
1938	1,431,567	10,390,000	11,821,567	24.9
1939	1,288,999	9,480,000	10,768,999	22.6
1940	1,094,252	8,120,000	9,214,252	19.3

* estimates from Historical Statistics of the United States; Unemployed, Part 1, p. 126, col. 8; and Discouraged workers, Current Population Survey data, Bureau of Labor Statistics, *http://www.bls.gov/webapps/ legacy/cpsatab13/htm*

These calculations and estimates are made because there was no discouraged worker data available for the era of the Great Depression. Personal income and the inverse of the rate of the total unemployed (number unemployed + discouraged workers) are shown in Figure 10.1. The number of persons in the adjusted civilian labor force (Table 10.2), discouraged workers, and adjusted unemployment rate (Table 10.2) are calculated.

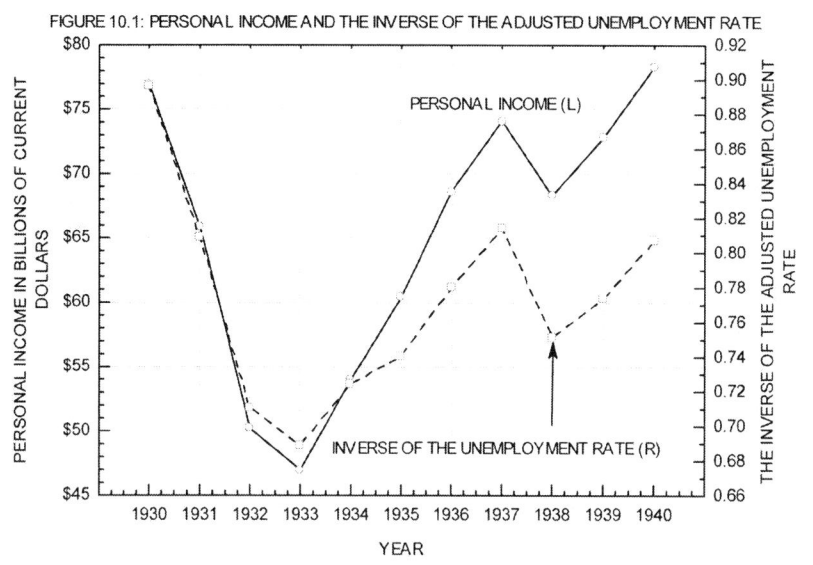

FIGURE 10.1: PERSONAL INCOME AND THE INVERSE OF THE ADJUSTED UNEMPLOYMENT RATE

SOURCE: CALCULATIONS FROM HISTORICAL STATISTICS OF THE UNITED STATES AND THE CURRENT POPULATION SURVEY (CPS)

Personal Income, p. 224, col. 8; Unemployment rate, Part 1, p. 126, col. 9, Discouraged workers, Table 13, http://www.bls.gov/webapps/legacy/cpsatab13.htm

his figure and the related calculations in Table 10.1 demonstrate a much more marked shift in the adjusted employment rate. This shift demonstrates a more accurate picture of the reality of the Great Depression era with unemployment peaking at 33% ((or a trough (inverse) of 67 %)) in 1933.

Further Hypotheses

8. Money supply (M_2 in billions of current dollars), personal income (PINC in billions of current dollars), the WPI, and the ADJUNR are related. M_2 may be used as a dependent variable. Personal income may also be used as a dependent variable.

9. The number of discouraged workers in the labor force and their inclusion in the unemployment rate provides a more accurate understanding of the severity of the Great Depression era (see Table 8.2 and Figure 8.1, Chapter VIII).

Hypothesis number eight will now be tested using two Equations. The first dependent variable tested will be M_2.

Table 10.3: Correlation Matrix (N = 41)

	ADJUNR	M_2	WPI
M_2	.42**		
WPI	NS	.49**	
PINC	NS	.95*	.69*

NS = not significant
**$p \geq .001$, *$p \leq .000$

Variable	Mean	Standard Deviation
ADJUNR	.085	.132
M_2	28.33	14.95
WPI	82.36	22.96
PINC	50.74	23.25

Equation 10.2

M_2 = .69 PINC -.14 WPI + 19.64 ADJUNR + 2.06   (10.2)
(t = 28.19) (t = -5.50) (t = 5.22)
(p < .000) (p < .000) (p < .000)

Adjusted R^2 = .98, F = 602.25, p = < .000, SEE =+ 2.20, N=41

Durbin-Watson = 1.98, p < .01, serial correlation = -.09

> **Equation 10.3**
>
> $$\text{PINC} = 1.39\, M_2 + .23\, \text{WPI} - 23.68\, \text{ADJUNR} - 4.02 \qquad (10.3)$$
> $(t = 28.19)\ (t = 7.51)\ (t = -4.02)$
> $(p < .000)\ (p < .000)\ (p < .000)$
>
> Adjusted $R^2 = .98$, $F = 720.77$, $p = <.000$, SEE $= +3.14$, N=41
>
> Durbin-Watson = 2.01, $p < .01$, serial correlation = -.08

The preceding Equation (Equation 10.3) uses personal income as a dependent variable.

What is to be summarized from Equations 10.2 and 10.3? Both dependent variables (M_2 money supply, and PINC or personal income) have a relationship with the WPI and ADJUNR ((adjusted unemployment rate, (number unemployed + discouraged workers)/(adjusted civilian labor force)) as well as a relationship with each other. These two Equations reveal the interdependence of PINC and M_2. Equation 10.2 suggests that PINC, the ADJUNR, and WPI need to be taken into consideration in the establishment of money supply levels. Equation 10.3, suggests that PINC is related to the three independent variables in varying degrees. The sign of the ADJUNR rate term is in the correct direction. M_2 represents an important part of the value of the Equation (10.3).

Equations 10.2 and 10.3 are suggestive of a role for M_2, PINC, the ADJUNR, and the WPI in the course of the economy. The signs of the independent variables are in the correct direction, and the significance of the Durbin-Watson statistics affirms the lack of autocorrelation in the Equations. The low serial correlations are acceptable.

It has already been determined that after 1933 there was an increase in M_2, personal income, and gold stocks. Figure 10.2 provides a graphic display of M_2, personal income, and gold stock.

FIGURE 10.2: M2, PERSONAL INCOME, AND GOLD STOCK

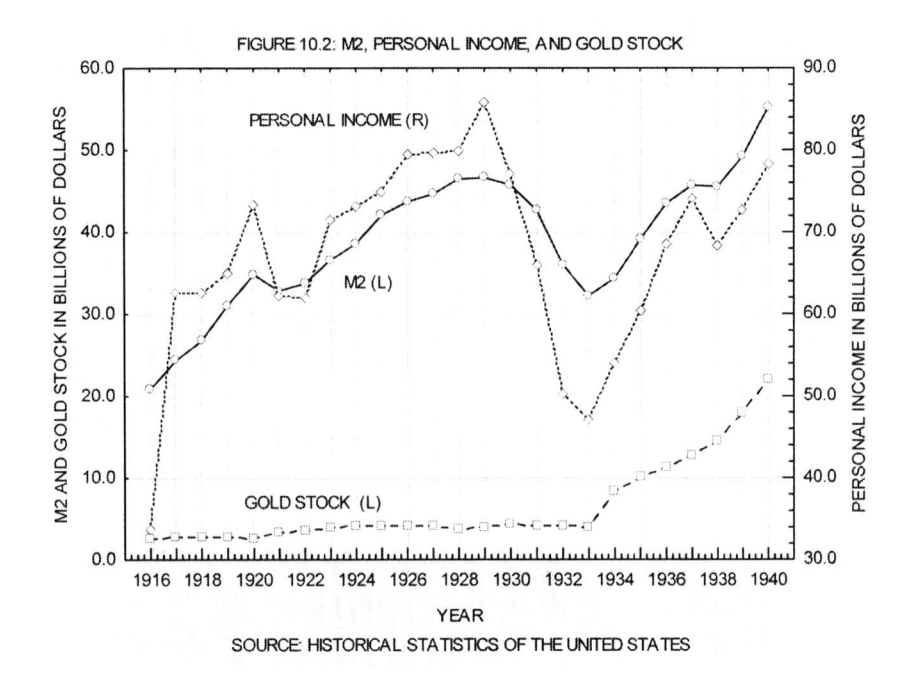

M₂, Part 2, p. 992, col. 415; Gold Stock, Part 2, p. 995. co. 438;
Personal Income, Part 1, p. 225 and Creamer, Daniel, p. 126, Table A-4

The joint decline of M_2 and personal income from 1929 through 1933, and their joint increase in value suggests that increasing money supply and personal income went hand in hand. Likewise, the formation of the FDIC, implemented in 1934, to protect the deposits of U.S. citizens, provided greater confidence in the banking system, making additional deposits safer. One must question the dip in M_2 from 1929 through 1933. Does it represent the (1) withdrawal of deposits and patrons from banks, (2) the hoarding of money, (3) the withdrawal of members of the labor force due to the decline of foreign-born males ages fifteen through seventy-four, (4) the large percentage of individuals who were unemployed and discouraged workers, (5) the failure to depart from the gold standard, or (6) the failure to implement the insurance of bank deposits? While the author tends to favor answering the question with a response of "all of the above," a certain answer to this question remains elusive. The empirical evidence in Equations 10.2 and 10.3

provides at least a partial response suggesting that M_2 and personal income are in an equilibrium relationship modified by the adjusted unemployment rate and the wholesale price index. (Note: In the Appendix II, an equation is developed for a period of 77 years, including the years of and beyond the Great Depression era.).

The next steps in the efforts to define our understandings of the depths of the Great Depression will take U.S. to examining the nature of (1) a conceptual framework for a disaggregated view of aggregate supply and aggregate demand and (2) the impact of immigration.

CHAPTER XI

LABOR FORCE AND AGGREGATE DEMAND

Adjustments to Labor Force Data

Because of the previously introduced argument that the statistics available for the depression era did not take into account discouraged workers, adjustments have been made to the number unemployed to create a statistic known as the total unemployed (number unemployed + discouraged workers). First, the original statistics from *Historical Statistics of the United States* will be presented again in Table 11.1. Table 11.2 contains the adjusted statistics.

Table 11.1: Labor Market Statistics, 1925-1940*

Year	Employment	Farm Employment	Non-farm Employment	Labor Force	Unemployed
1925	43,716,000	10,662,000	33,054,000	45,431,000	1,453,000
1926	44,828,000	10,690,000	34,138,000	45,885,000	801,000
1927	44,856,000	10,529,000	34,327000	46,634,000	1,519,000
1928	45,123,000	10,497,000	34,626,000	47,367,000	1,982,000
1929	46,207,000	10,541,000	35,666,000	48,017,000	1,550,000
1930	44,183,000	10,340,000	33,843,000	48,783,000	4,340,000
1931	41,305,000	10,240,000	31,065,000	49,485,000	8,020,000
1932	38,038,000	10,120,000	27,918,000	50,348,000	12,060,000
1933	38,052,000	10,090,000	27,962,000	51,132,000	12,830,000
1934	40,310,000	9,990,000	30,020,000	51,910,000	11,340,000
1935	41,673,000	10,110,000	31,563,000	52,553,000	10,610,000

1936	43,989,000	10,090,000	33,899,000	53,319,000	9,030,000
1937	46,608,000	10,000,000	36,068,000	54,088,000	7,700,000
1938	44,142,000	9,840,000	34,302,000	54,872,000	10,390,000
1939	45,738,000	9,710,000	36,028,000	55,588,000	9,480,000
1940	47,520,000	9,540,000	37,980,000	56,180,000	8,120,000

* Historical Statistics of the United States, Part 1, p. 126, col. 5, 6, 7, 8

Table 11.2: Adjusted Civilian Labor Force (CLF), the Adjusted Unemployed, the Number Unemployed, and the Number of Discouraged Workers*

Year	CLF Adjusted (in billions) (1)	Adjusted Unemployed (in millions) (2)	Number Unemployed (in millions) (3)	Number of Discouraged Workers (2)-(3)
1930	48.22	5.01	4.34	666,890
1931	48.34	9.24	8.02	1,220,252
1932	47.89	13.87	12.06	1,811,287
1933	47.35	14.73	12.83	1,897,702
1934	47.03	12.99	11.34	1,651,121
1935	46.79	12.14	10.61	1,525,395
1936	46.94	10.31	9.03	1,278,659
1937	47.28	8.77	7.70	1,073,598
1938	47.45	11.82	10.39	1,431,567
1939	47.69	10.77	9.48	1,288,999
1940	47.85	9.21	8.12	1,094,252

* Source: Col. 1—CLF, Part 1, p. 126, col. 4; Col. 2—Calculations from Current Population Survey data, Bureau of Labor Statistics and col. 3, http://www.bls.gov/webapps/legacy/cpsatab13/htm; Col. 3—Part 1, p. 126, col. 8

Table 11.3: The Adjusted Civilian Labor Force, Adjusted Number Of Unemployed, and the Adjusted Unemployment Rate

Year	Civilian Labor Force (Adjusted)	Adjusted Unemployed (in millions)	Unemployment Rate (Adjusted)
1930	48,219,523	5,006,890	10.4%
1931	48,338,676	9,240,252	19.1
1932	47,893,277	13,871,287	29.0
1933	47,349,666	14,727,702	31.1
1934	47,028,956	12,991,121	27.6
1935	46,791,671	12,135,395	25.9
1936	46,937,242	10,308,659	22.0
1937	47,283,522	8,773,598	18.6
1938	47,446,855	11,821,567	24.9
1939	47,685,767	10,768,999	22.6
1940	47,850,644	9,214,252	19.3

Source: Calculations from Current Population Survey data; Bureau of Labor Statistics,

Column (1) the adjusted civilian labor force is equal to the civilian labor force minus the annual cumulative decline in the number of foreign-born male workers ages 15 through 74 from 1930 through 1940. The adjusted unemployed (Column 2) is equivalent to the number unemployed plus discouraged workers. Use of the adjusted number of unemployed (the unemployed + discouraged workers) divided by the adjusted civilian labor force in calculating an unemployment rate yields the following results (see Table 11.3):

Usual estimates of the unemployment rate are conservative for at least two reasons. First, as previously addressed, the estimate of the unemployed is deflated as it is an interpolation between census data for 1930 and 1940 and did not take into account the working age males and females ages 15 through 74 who were discouraged workers. This underestimate of the unemployed labor force results in a smaller unemployment rate. Second, an estimate of the number of discouraged workers using *Current Population Survey* data and *Historical Statistics of the United States* data with the unemployment rate as an independent variable was made. The adjusted unemployment estimate combining the number unemployed and the number of discouraged workers provides a more accurate perspective on unemployment in the Great Depression era. These processes resulted in a mean adjusted unemployment averaging 10,805,429 for the 1930 through 1940 period (N = 11, standard deviation = 2,744,876). The mean of the adjusted civilian labor force was 47,529,594 with a standard deviation of 513,400 (N = 11).

Using the data for the adjusted number unemployed, its value (14.728 million) peaked in 1933 along with that of the adjusted unemployment rate (31.1%). This coincides with other data that support the view that the most severe year of the Great Depression era was 1933.

Aggregate Data

The following aggregate data from the National Income and Product Accounts (NIPA) (Bureau of Labor Statistics, Department of Commerce) (Table 11.4) are presented and will be analyzed here and in the Appendix II.

Table 11.4 provides an accurate overview of the persistence of the Great Depression era. Aggregate demand remained below its 1929 levels for the entire 1930 through 1940 period. This characterization of the Great Depression era is also true of gross domestic private investment, and personal consumption. The decline in aggregate demand from 1929 through 1933 was $76 billion or 73%. The decline in personal consumption (limited by personal income) from 1929 through 1933 was $31.5 billion or 68.3%. These declines indicated the seriousness and depths of the Great Depression era.

Table 11.4: Personal Consumption, Government Capital Expenditures, Gross Domestic Private Investment, Net Exports, and Aggregate Demand (in Billions of Current Dollars)*

Year	Personal Consumption	Government Capital Expenditures	Gross Domestic Private Investment	Net Exports	Aggregate Demand
1929	77.4	9.4	16.5	.4	103.7
1930	70.1	10.0	10.8	.3	91.2
1931	60.7	9.9	5.9	1.0	76.5
1932	48.7	8.7	1.3	0	58.7
1933	45.9	8.7	1.7	.1	56.4
1934	51.5	10.5	3.7	.3	66.0
1935	55.9	10.9	3.7	-.2	73.3
1936	62.2	13.1	8.6	-.1	83.8
1937	66.8	12.8	12.2	.1	91.9
1938	64.3	13.8	7.1	1.0	86.2
1939	67.2	14.8	9.3	.8	92.1
1940	71.3	15.0	13.6	1.5	101.4

* Department of Commerce, Bureau of Economic Analysis; http://www.bea.gov/bea/dn/nipaweb/, list of all NIPA Tables

Recovery was slow in developing. From 1933 through 1940, aggregate demand increased by $45 billion or 80%. This was an average annual increase of 11 percent. Similarly, personal income increased $31.6 billion or an average annual amount of 67%. Government capital expenditures ($6.3 billion) and gross domestic private investment ($11.9 billion) experienced more rapid rates (72% and 700% respectively) of recovery than did personal income. Net exports made little contribution to the nascent recovery.

The relationship between aggregate demand (dependent variable) and its component independent variables (personal consumption, government capital expenditures, gross domestic private investment, and net exports) is a straightforward mathematical equality.

Aggregate Demand	Personal Consumption	Government Capital Expenditures	Gross Domestic Private Investment	Net Exports	
$AD_{xt} =$	PC_{xt} +	GOV_{xt} +	$GDPI_{xt}$ +	Z_{xt}	(11.1)

Thus, it is always an Equation with 100% of the variation in the dependent variable explained. All variables in the aggregate demand Equation are correlated above .96 (p = 0.00, N = 77, 1929 through 2005) with the other variables making up the aggregate demand Equation with the exception of net exports. Net exports correlate negatively between a range of -.84 and -.86 (p = 0.00, N = 77) with the other variables in the aggregate demand Equation.

Equations 7.3 and 7.4 (Chapter VII) have some major lessons to teach us: First, aggregate demand can be disaggregated into its component parts: the four independent variables plus their component parts. Secondly, by disaggregating these variables, internal variation in each of them may be found that significantly affects each of the variables and in the end the term for aggregate demand (AD). This is particularly true of the term for personal consumption. Personal income and personal consumption are highly related.

The variation in the term for personal income, and therefore for personal consumption, depends on conditions set forth in Chapter X. These theses (in positive form) partially repeated here are:

1. The variation of the number of individuals (of income-earning age) by specified subgroups that differ significantly on average personal income affects total real personal income for the subgroup and therefore personal consumption in a developed economy. This variation has its source in the number of individuals born in each subgroup, the number passing away in each subgroup, the net immigration into and out of each subgroup, the participation rate in income economy, and the sex, ethnicity, age, and average income.

2. There is variation in the *rates* of individuals in each subgroup who have personal income. This rate, the personal income

participation rate, or the employment rate for each subgroup, affects personal consumption and, therefore aggregate demand.

3. Differences in *real personal income by subgroup* affect total real personal income for each subgroup and, therefore total personal income for the population.

4. Changes in subgroup real personal income may act as a leading influence on economic growth and decline as measured by changes in personal consumption and aggregate demand. Also affected are changes in levels of savings (and debt), aggregate demand, aggregate supply, changing output, and changing employment and unemployment. A multiplier effect may also occur from changes in personal income.

5. Wage and salary levels of individuals are "sticky." Thus, periods of economic decline are generally not associated with major downward shifts in wages. Long periods of economic decline, such as depressions, periods in which there are large, relative influxes of immigrants or nonworking poor into the labor market may put more significant downward pressures on domestic wages.

6. Stress Theory provides a theoretical and applied approach for evaluating both equilibrium and "stress" for different components of the economy such as aggregate supply and aggregate demand. Stress Theory is particularly useful for evaluating what Fisher (1911) termed "transition" periods, or periods of asymmetry, empirically.

7. The terms for personal income, government capital expenditure, gross domestic private investment, and net exports also can be disaggregated. Their disaggregated variation also affects aggregate demand and aggregate supply.

Gross National Product

To demonstrate the relationship of personal income, M_2, and the inverse of the adjusted unemployment rate, the adjusted unemployment rate was inverted (1-unemployment rate). Figure 11.1 shows the personal income, M_2, and the inverse of the adjusted unemployment rate declining to lows (highs for the unemployment rate) in 1932-1933. There was a secondary recessionary period in 1938. Figure 11.1 traces three variables (inverse of the adjusted unemployment rate, personal income, and M_2). The three variables decline from 1929-1933 and then rebound. These variables reach simultaneous lows in 1933.

Consumer Price Index

One of the economically distinctive features of the Great Depression was a persistent deflation (using 1926 as a comparison year). From a peak of 103.5 (1926 =100) in 1925 the WPI declined to a low of 64.8 in 1932. This represented an increase in the real value of the dollar as well as adjustment to a lack of demand. The WPI increased from its low in 1933 to a high of 86.3 in 1937. The 1929-1932 period reflected a time of declining prices, declining demand, and a declining money supply. It also reflected a declining confidence of the public in their financial institutions. Despite this increase in the real value of the dollar, current personal income and personal consumption declined from 1929 through 1933. After 1933, there was an increase in the two quantities that remained short of 1929 levels.

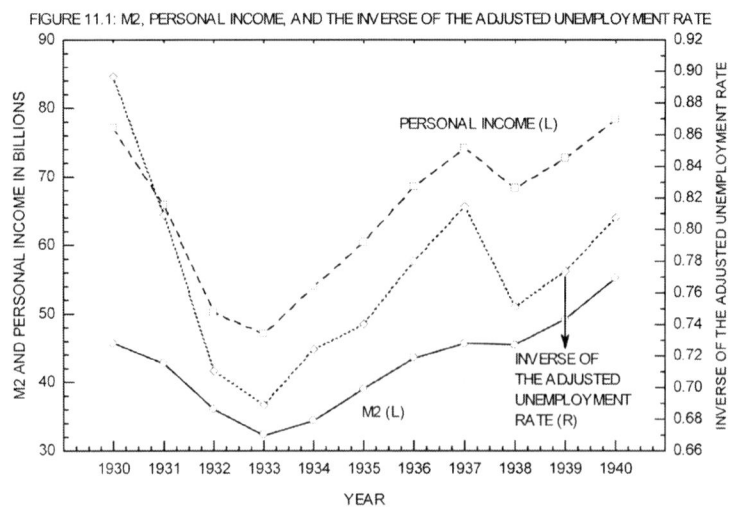

FIGURE 11.1: M2, PERSONAL INCOME, AND THE INVERSE OF THE ADJUSTED UNEMPLOYMENT RATE

Source: Historical Statistics of the United States and calculations from the Current Population Survey

Personal income, http://www.bea.gov/bea/dn/nipaweb/, list of all NIPA Tables; M_2, Historical Statistics of the United States, Part 2, p. 992, col. 415; Inverse of the Adjusted Unemployment Rate, calculated based on BLS data.

The declining adjusted unemployment rate (in inverse form) with a declining level of personal income in the period from 1929 through 1933. The relationships in Figure 11.6 are further elucidated by the correlations of the three variables. Table 11.5 presents their correlations with an additional variable, the wholesale price index (WPI) (N = 41).

Table 11.5: Correlation Matrix (N=41)

	Inverse of Unemployment Rate	M_2	Personal Income
M_2	-.42*	—	
Personal Income	NS	.95**	—
CPI	NS	.84*	.93**

*p = ≥ .001, ** p ≤ .000

The WPI was added because of its strong relationship with personal income and its lesser relationship with M_2. At first examination, the decline in personal income appears to go hand-in-hand with a decreasing WPI. Demand declines, and business and industry respond to the decline in personal income by cutting prices. If decreasing prices are not successful, decreasing production follows. It is not appropriate to infer causation from any of these relationships.

Magnitude of Impact on Personal Income

The magnitude of the statistics for personal income were based on Bureau of Labor Statistics NIPA tables from 1930 through 1940. Table 11.6 presents the adjusted number unemployed and the personal income decline with the annual percent change.

Table 11.6: Adjusted Unemployed, Personal Income (in Billions), and Percent Change in Personal Income

Year	Adjusted Unemployed	Personal Income (in billions)	Percent change in Personal Income
1930	5,006,890	$77.0	-11.56%
1931	9,240,252	65.9	-16.84
1932	13,871,287	50.2	-31.27
1933	14,727,702	47.0	-6.81
1934	12,991,121	54.0	12.96
1935	12,135,395	60.4	10.60
1936	10,308,659	68.6	11.65
1937	8,773,598	74.1	7.42
1938	11,821,567	68.3	-8.49
19399	10,768,999	72.9	6.18
1940	9,214,252	78.3	7.02

Source: Calculations from Current Population Survey data, Campbell and Lennon, and *http://www.bea.gov/bea/dn/nipaweb/* list of all NIPA Tables

Finding Quantitative Guidelines

The Federal Reserve has searched for quantitative guidelines for regulating the U.S. economy. The lack of success is marked by a need to correctly estimate velocity. The regression Equations (Chapter IX, 9.2 and 9.3) appear to be adequate for the 1900 through 1940 period as they both indicate through their Durbin-Watson statistics that autocorrelation is not significant. Both Equations 9.2 and 9.3 indicate that both the adjusted unemployment rate and the WPI have some influence on M_2 and personal income.

As the discussion of the Great Depression is brought to a close, there are parallels as well as differences between the Great Depression and the current economic crisis. The relationship between mean income of working persons, their ages, sex, and numbers is found to be an essential component of any analysis of economic recessions and depressions.

CHAPTER XII

EXPLAINING THE GREAT DEPRESSION

Immigration

Immigration was an important factor in the period from 1900 through 1940. To support this view a series of figures starting with Figure 12.1 will be presented and analyzed.

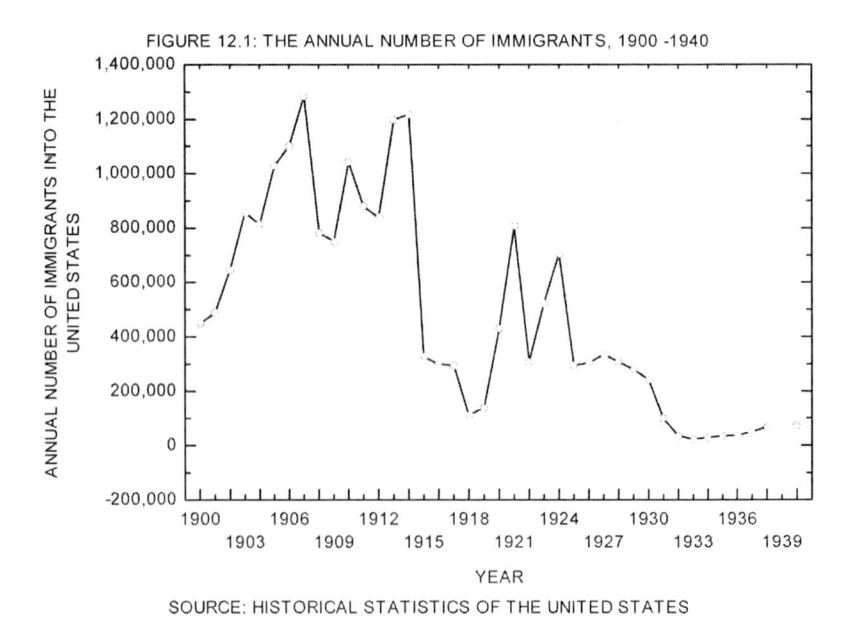

FIGURE 12.1: THE ANNUAL NUMBER OF IMMIGRANTS, 1900 -1940

SOURCE: HISTORICAL STATISTICS OF THE UNITED STATES

*Source: Historical Statistics of the United States, Part 1, p. 105, col. 89

From 1900 to 1914 there were 13.38 million immigrants into the U.S. Of these immigrants, 69.2 percent, or 9.25 million were males. Immigration declined rapidly to 1.17 million individuals in 1915 to 1919. There were peaks in 1921 (805,228 immigrants) and 1924 (706,896 immigrants). Immigration dropped to 241,700 in 1930. From 1931 to 1940, annual immigration never exceeded 100,000 individuals.

What remains important about the magnitude of the number of immigrants from 1900 to 1940 is their impact on the United States economy. Two of these impacts will be presented graphically and then analyzed using qualitative and quantitative analysis.

Live Births and Immigration

Live births need to grow to working age to have a major impact on the economy. As the live births age, they, in different groups and different proportions, become part of the producing, income-earning, and consuming economy.

The number of immigrants lagged 10 years is correlated with the number of live births at $r = .57$, $p=.000$, $N=41$. The drop in the number of immigrants from 1914 to 1915 lagged 15-16 years is the underlying population dynamic partially responsible for the Great Depression. That population dynamic represents the declining number of persons averaging 48-49 years old. Put in other terms, there is a distribution of persons 48-49 years of age, aging through the Depression years. Caution should be taken in interpreting these numbers as the number of individuals around the 48-49 lag would preferably be multiplied by their survival rate.

FIGURE 12.2: LIVE BIRTHS COMPARED TO ALL IMMIGRANTS, 1900-1940

SOURCES: HISTORICAL STATISTICS OF THE UNITED STATES AND SOCIAL SECURITY

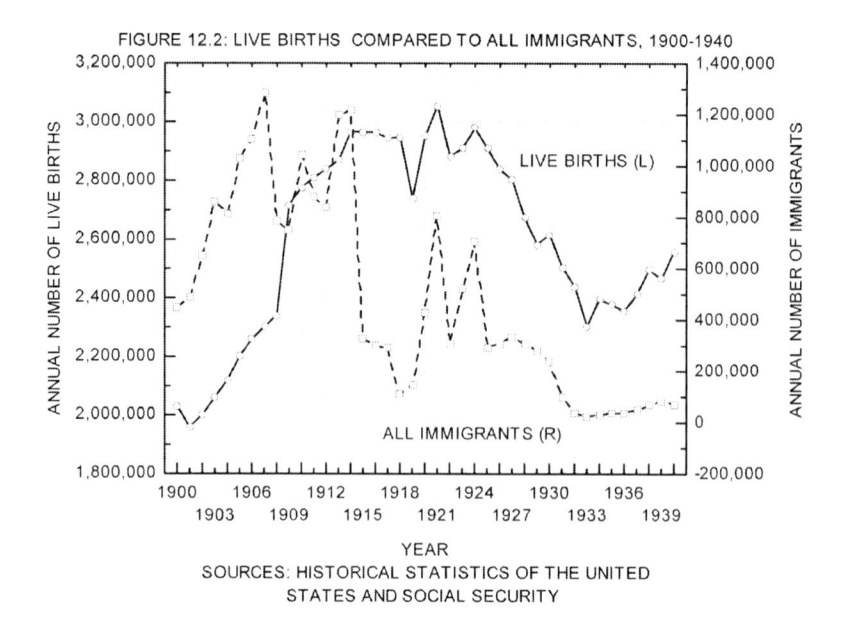

Social Security, National Center for Health Statistics, and Historical Statistics of the United States, Part 1, p. 105, col. 89

All annual points on the live birth line represent the number of individuals of zero years of age or those in their first year of birth. All annual points of immigrants on the immigrant line represent the number of individuals distributed around an average age of an estimated 33-34 years.

If live births are lagged 16 years, just as immigrants are, Figure 12.3 is the result. Live births lagged 16 years (and adjusted for survival) start entering the population in 1916 as working individuals. Live births lagged sixteen years from 1929 to 1940 (35.17 million) and adjusted for survival rates enter the period of the Depression. It is likely that these males and females earned low wages because they were entering the labor force. Immigrants lagged 16 years were correlated with live births lagged 16 years r = -.43 (p=.03, N=25).

This provides evidence that immigrants lagged 16 years played a larger role in the contraction than did the influx of live births 16 years of age because the live births did not have sufficient opportunity to grow to working age. This information was supplemented by Figure 12.4 which tests the relationship of the variable immigrants lagged 16 years to the inverse of the adjusted unemployment rate. These two variables (1916-1940) were related at r = .72, p = .000, N= 24.

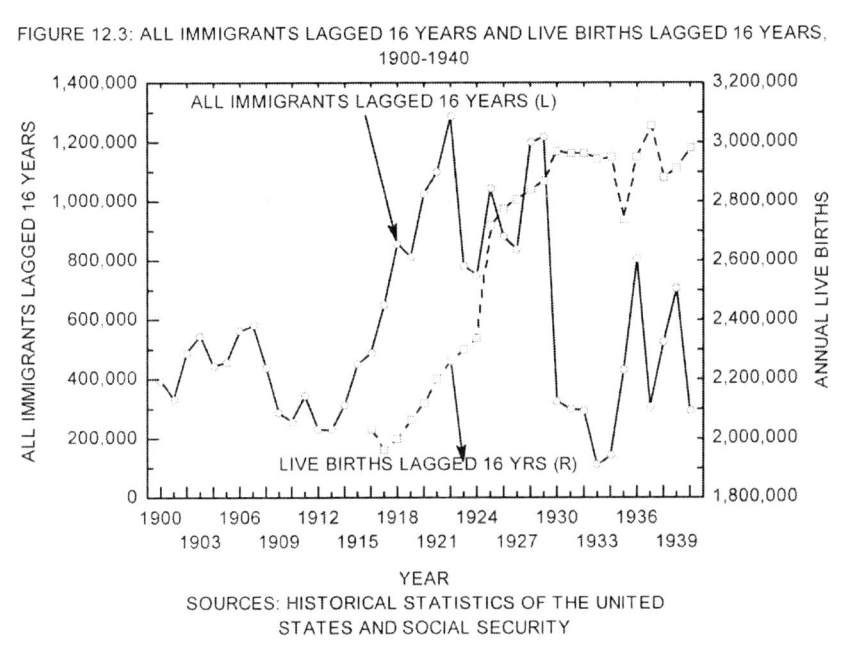

FIGURE 12.3: ALL IMMIGRANTS LAGGED 16 YEARS AND LIVE BIRTHS LAGGED 16 YEARS, 1900-1940

SOURCES: HISTORICAL STATISTICS OF THE UNITED STATES AND SOCIAL SECURITY

Historical Statistics of the United States, Part I, p. 105, col. 89, http://www.ssa.gov/OACT/NOTES/pdf_studies/study120.pdf, and http://cdc.gov/nchs/data/statab//4941x.01.pdf

From 1929-1930 through 1933 an increase in unemployment was accompanied by a decline in immigration. While correlation statistics cannot be used to draw causal inferences, Figure 12.4 raises questions about the impact of immigration. If the average age of immigrants in the 1900-1914 period was 34 years, a lag of 16 years would make the immigrants' lagged distribution around an average age of 50 years.

Housing: A Prelude to U.S. Crashes?

The Great Crash as well as the current economic crisis are preceded by a housing bubble. See Figure 12.5. The fact the numbers of housing starts from 1920-1940 and from 1988-2008 were different in magnitude, and the fact that both housing bubbles came prior to economic crises hints that a common variable has been in operation. There were 10.21 million housing starts in the 1920-1940 period. There were approximately 14 million immigrants lagged 16 years during that period. It would be unjustified to attribute the totality of housing totally to the lagged immigrants, but later Equations will provide some clarification.

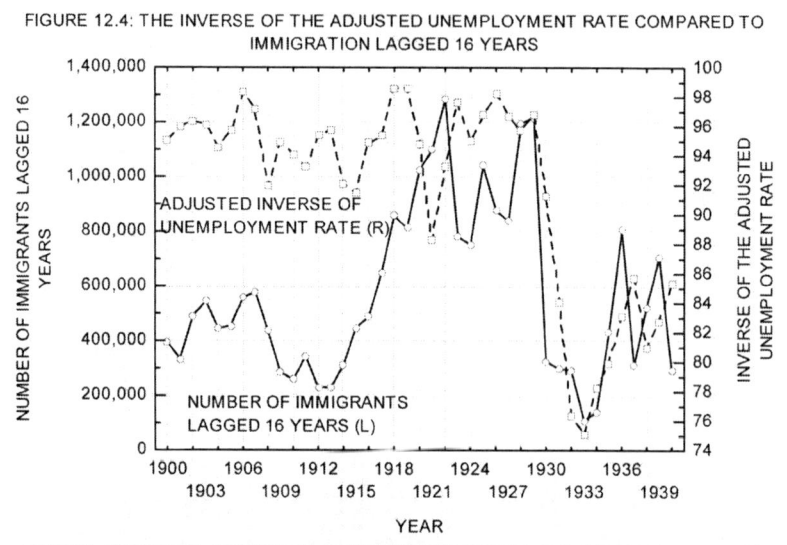

FIGURE 12.4: THE INVERSE OF THE ADJUSTED UNEMPLOYMENT RATE COMPARED TO IMMIGRATION LAGGED 16 YEARS

SOURCE: HISTORICAL STATISTICS OF THE UNITED STATES AND CALCULATIONS FROM BUREAU OF LABOR STATISTICS DATA, r=.70, p=.001, N=18

Calculations using the unemployment rate and the number of immigrants

THE CURRENT ECONOMIC CRISIS AND THE GREAT DEPRESSION   221

FIGURE 12.5: ALL IMMIGRANTS LAGGED 16 YEARS VS. PRIVATE HOUSING STARTS

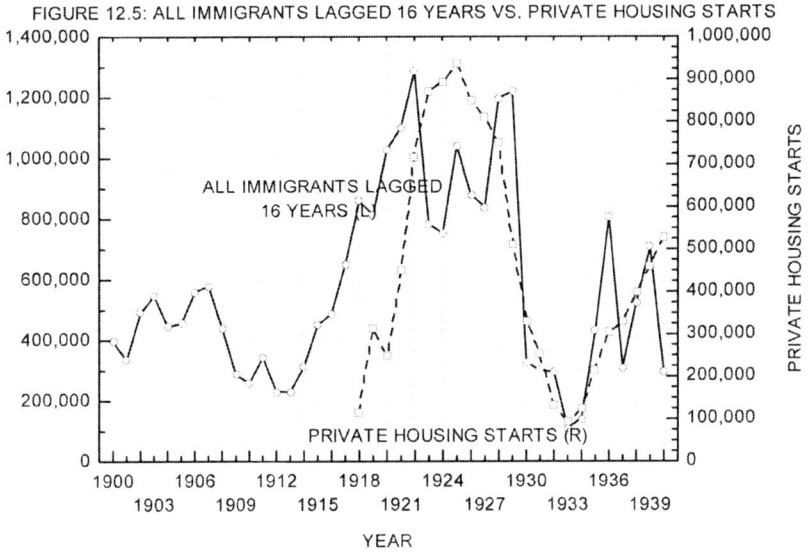

SOURCE: HISTORICAL STATISTICS OF THE UNITED STATES, VOL. 1 & VOL. 2, r = .63, p = .002, N = 21

Part I, p. 105, col. 89 and p.

In the 1921-1925 period a housing bubble developed. Private housing starts increase from 247,000 houses in 1920 to 937,000 in 1925. From 1925 through 1928 the rate of decline from the 1925 peak slowed. There was a much more rapid decline from 1929 to 1933 (from 753,000 private housing starts to 93,000). Banks were loaning money in the 1921-1928 period without the rigor required in approving loans. Nonfarm foreclosures peaked in 1932 and 1933 (at 252,400), leveled during 1934 and 1935, and started a regular decline lasting until 1945 (12,706). Housing starts marked a bottom in 1933 (93,000) and increased at a steady rate from 1934 to 1940. Nonfarm foreclosures in the 1934-1940 period declined slowly at first (1934-1935) and continued a more rapid decline up to 1945.

As can be seen from Figure 12.6, the number of immigrants lagged 16 years coincides with the housing bubble and vice versa. Assuming that the immigrants ages fell in a range surrounding 34 years of age upon their arrival into the U.S., lagging them 16 years makes for an age range centered on 49-51 years.

Of importance to the private housing starts is the demand due to the age distribution of home purchasers. The only data for such demographics that the author has found are for the 1994-2003 period. Another comparison indicates that homeowners would tend to be concentrated in the 30 and above age range and thus would be part of the distribution of immigrants lagged 16 years. While the comparison to the 1994 to 2003 period data is suspect, it is likely that the home—buying age in the 1920-1940 period was lower than that of the later period from which the following data come.

Table 12.1: Age and Percent Owning Homes, 1994-2003*

Age of Homeowners	Percent Owning Homes
<25	19.4%
25-29	36.6
30-34	53.7
35-44	66.8
45-54	76.0
55-64	80.5
65&over	78.5

*Source: www.realclearmarkets.com

The variables, immigrants lagged 16 years (IMMIGL16) and private housing starts (PHOUSESTAR) reach a low point in 1933. Additionally, nonfarm foreclosures (NFARMFORCL) reached a peak in 1933 (see Figure 12.6). What follows is a correlation matrix with a small N, for the variables private housing starts, immigrants lagged 16 years, the inverse of the adjusted unemployment rate, and non-farm foreclosures.

Table 12.2 Correlation Matrix (N=15)

	All Immigrants Lagged 16 years	Inverse of the Adjusted Unemployment Rate	Non-farm Foreclosures
Inverse of the Adjusted Unemployment Rate	.76*	—	
Non-farm Foreclosures	-.56*	-.74	—
Private Housing Starts	.75**	.88***	.88***

*p = < .05, **p = ≤ .002, ***p = .000

FIGURE 12.6: IMMIGRANTS LAGGED 16 YEARS, PRIVATE HOUSIING STARTS, AND NONFARM FORECLOSURES

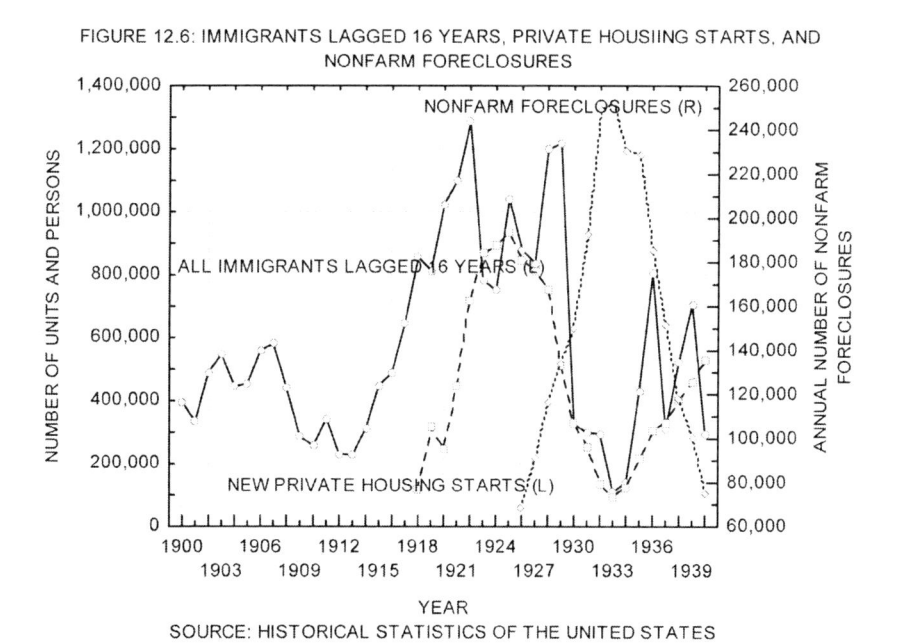

SOURCE: HISTORICAL STATISTICS OF THE UNITED STATES

Part I, p. 112, col. 138; Part II, p. 651, col. 301, Part II, p. 640, col. 158

The Great Depression was preceded by an expansion of credit (loose credit). When the Crash hit, credit suddenly became difficult to find. A lack of trust pervaded relationships between borrowers and lenders. Following the Crash from 1930 to 1932-1935 nonfarm closures increased to about 250,000. In 1933 private housing starts declined to approximately 93,000.

To confirm the congruence of immigrants lagged 16 years with private housing costs, Figure 12.7 was prepared. Visually, the congruence of the peaks is similar to that for Figure 12.6.

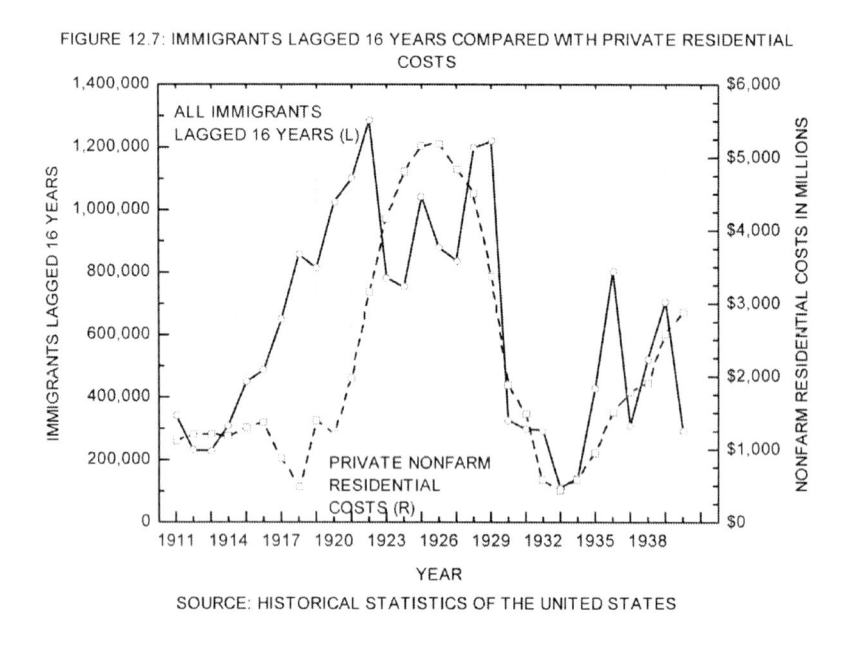

FIGURE 12.7: IMMIGRANTS LAGGED 16 YEARS COMPARED WITH PRIVATE RESIDENTIAL COSTS

SOURCE: HISTORICAL STATISTICS OF THE UNITED STATES

Part II, p. 623, col. 72

The peak in nonfarm foreclosures (1933) followed the peak in private housing starts (1925) by eight years. This pattern leads to the question; will foreclosures in the current economic crisis follow the peak in housing starts in 2004 with a pattern similar to the pre-Depression through Depression period?

What does this housing start and foreclosure information indicate? Several alternatives will be presented.

1. The foreclosure bubble indicates that banks granted loans to individuals who were not credit-worthy. There a period of very loose credit from 1922 to 1929.

2. There was a tightening of credit from 1929 to 1933 with a gradual loosening of credit from 1934 to 1940.

3. The source of this credit boom, and bust was the housing crisis, the foreclosure crisis, and the lack of faith of depositors in U.S. banks. Easy credit was available for traders on the stock market up to the 1929 Crash.

4. The dip in M_2 from 1930-1933 was a consequence of a contraction in personal income and therefore deposits of checkbook money as well as a lesser decline in M_1.

Kindelberger (1989) would say that one of the initial stages typical of manias, panics, and Crashes is a loosening of credit. This appears to have been what happened in the stage before the Depression. The foreign-born white subgroup of those who immigrated from 1900 to 1914 (estimated average 33-34 years of age on arrival into the U.S.), lagged sixteen years, would on average, be 49-50 years of age during the 1916-1929 period. The dramatic drop of immigrant population from 1914 to 1915 is reflected in the rapid decline in immigrants lagged 16 years (1929-1930). The fourth alternative is best answered by Kindelberger's fifth stage. Credit becomes tight as there is a reinforcement of tighter standards for loans. Foreclosures, business failures, and high levels of unemployment make credit-worthy individuals and businesses more difficult for banks to find.

A second factor to be considered is the life expectancy of the immigrants in the 1900-1914 period. During that period approximately 69.2 percent of all immigrants were males. The earliest life tables available are for males and females born in 1900 (Bell and Miller, 2009). Approximately .6382 of the males and .68247 of the females would have survived until the 1916-1929 period. As the large number of immigrants from the 1900-1914 period (13.38 million persons) moved through the 1920-1940 period, aged, became unemployed, and reached their age of mortality, they had an economic impact.

A third and very important factor to consider is personal income. The variation in personal income in the 1900-1940 period is important evidence for the events of that period. There are two Equations of practically equal import that indicate the relative role of immigrants and the unemployed. The two Equations follow with their attendant statistics.

Table 12.3 Correlation Matrix (N=41)

	Reserves	Immigrants Lagged 16 Years	Total Unemployed	M_2	Wholesale Price Index
Immigrants Lagged 16 Years	.359*	—			
Total Unemployed	.521***	-.344*	—		
M_2	.997***	.386*	.502**	—	
Wholesale Price Index	.471**	.651**	-.18 NS	.488**	—
Personal Income	.948***	.543**	.291 NS	.955***	.687***

*p = < .05, **p = ≤ .002, ***p = .00

Table 12.3 Correlation Matrix (N=41)

	Reserves	Immigrants (Lagged 16 years)	Total Unemployed	M_2 Supply	Wholesale Price Index
Immigrants (Lagged 16 years)	.359 (p=.021)	—			
Total Unemployed	.521 (p=.000)	-.344 (p=.029)	—		

M_2	.997 p=.00	.386 (p=.029)	.502 (p=.001)	—	
Wholesale Price Index	.471 (p=.002)	.651 (p=.001)	-.18 NS	.488 (p=.001)	—
Personal Income	.948 (p=.00)	.543 (p=.00)	.291 (p=.064)	.955 (p=.00)	.687 (p=.00)

Standard Deviations and Means (N=41)

	Standard Deviation	$\overline{X}$
All Immigrants Lagged 16 Years	314,721	573,755
Total Number Employed	6,210,119	6,438,135
M_2	14.9	28.3
Wholesale Price Index	23.0	82.4
Personal Income	23.0	50.7
Bank Reserves	21.9	44.1

Equation 12.1: Dependent Variable Personal Income (N=41)

PI = .00001 IMMIGL16 + 1.256 M2 + .250 WPI -8.432
  (t=2.22)   (t=29.19)   (t=7.34)   (t=-3.99)
  (p=.032)   (p=.000)   (p=.000)   (p=.000)

Adjusted R^2 = .976, Durbin-Watson = 1.65, S.C.=.163, F= 566.08, p= .000, SEE = +3.53

PI = personal income (in current dollars and billions)

IMMIGL16 = The number of immigrants lagged 16 years

M_2 = M_2 money supply in billions

RESERVES = Banking reserves in billions

WPI = Wholesale price index

Equation 12.2: Dependent Variable Reserves (N=41)

RESERVES = .000001 TOTUNEMP + 1.04 PI - .202 WPI - 6.947
     (6.34)   (t=27.38) (t=-5.72) (t=-3.183)
     (p=.000) (p=.000) (p=.000) (p=.0006)

Adjusted R^2 = .98, Durbin-Watson = 2.06, S.C.= - 0.049, F= 650.7, p= .000, SEE = 3.10

TOTUNEMP = Total number of the unemployed

The proximity of these two equations and the previous analysis suggest that equation 12.1 and equation 12.2 have a closeness that is not accidental. The signs are correct.

Beta Weights			
Equation 1	Beta	Equation 2	Beta
IMMIGL16	.07	TOTUNEMP	.18
M_2	.81	PINC	1.04
WPI	.25	WPI	-2.12

Equation 12.2 is a successful Equation with a serial correlation that is smaller than that of Equation 12.1 (-.049||<1.63). The standard errors of estimate are slightly larger for Equation 12.1 than for Equation 12.2. The F value for equation 12.2 is larger than the F value for equation 12.1. When

the variable IMMIGL16 is replaced by the total number unemployed (and possibly included in the TOTUNEMP (under the principal of last hired, first fired) equation 12.2 remains the strongest equation.

The aging (movement) of the immigrants through the U.S. economy appears to have a defined impact on the economy through the decline in the number of employed individuals, their mortality, and changing wage rates. Unlike the analysis of the current economic crisis, an approximate distribution of income for the Great Depression is not available for analysis. Equations 1 and 2 do, however, allow a comparison of the relative impact of their contributing variables by analysis of their **Beta weights.** These follow:

Equation1	βeta	Equation2	βeta
IMMIGL16	.071	TOTUNEMP	.178
M_2	.807	PINC	1.04
WPI	.25	WPI	-2.12

The *beta coefficients* are the regression coefficients obtained by first standardizing all of the variables to a mean of zero (0) and a standardized coefficient of one (1). These coefficients allow the relative contribution of each independent (predictor) variable to a dependent variable to be assessed.

An important aspect of Equations 12.1 and 12.2 is the fact that the dependent variable differs. RESERVES, the dependent variable in Equation 12.2, is an indicator of banking activity. Banking activity during the 1930-1933 period varied greatly due to changes in personal income and effective demand. The level of personal income is highly correlated with personal consumption (r=.99, p=.000, N=77). Personal consumption represents 68 to 70 percent of economic activity. The adjusted unemployment rate translates into the number of unemployed (looking for work) plus the number of discouraged workers (employable, but not looking for work) divided by the number in the civilian labor force. When personal income declines or increases, the National Bureau of Economic Research Dating Committee watches it as an indicator of a start or end of a recession.

The presence of the RESERVES variable as a more explanatory variable than M_2 requires discussion, particularly because of runs on banks and the practice of hoarding during the Great Depression.

Table 12.4: Banking Reserves in the U.S.: 1930-1940

Year	Amount (billions)	Year	Amount (billions)
1930	$74.29	1936	$66.85
1931	70.07	1937	68.40
1932	57.30	1938	67.73
1933	51.36	1939	73.19
1934	55.91	1940	79.73
1935	59.91		

Path Diagram, Equation 1:

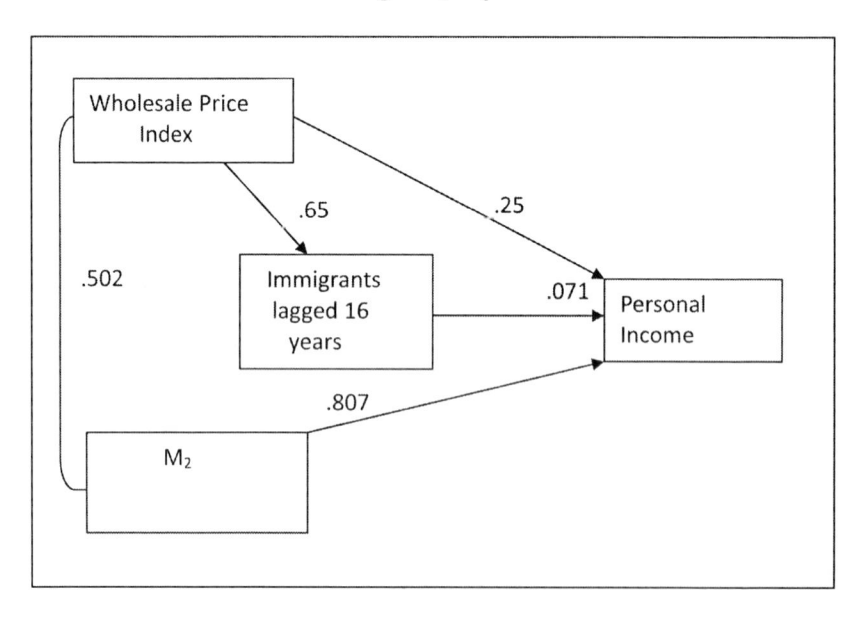

```
Wholesale Price Index ──.65──> Immigrants lagged 16 years ──.071──> Personal Income
                    \──.25────────────────────────────────────────>
.502
M₂ ──.807──> Personal Income
```

Path Diagram, Equation 2:

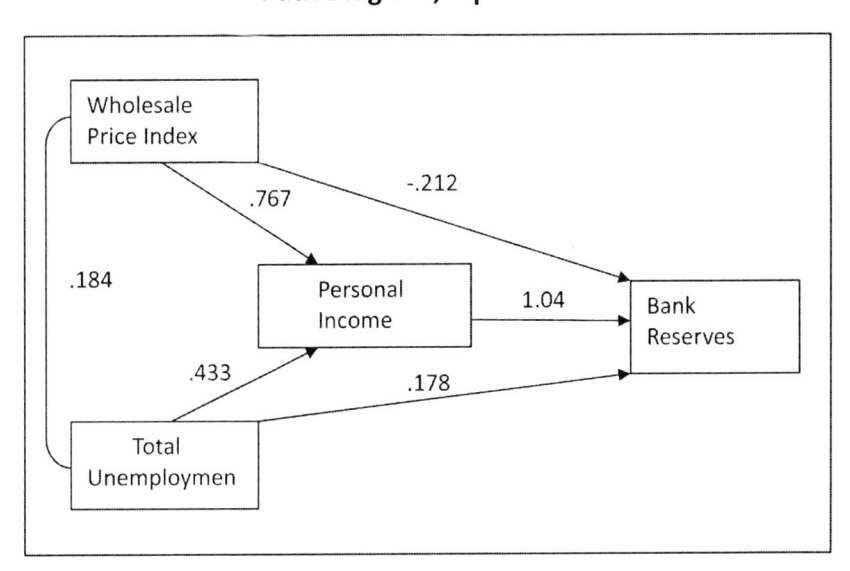

```
Wholesale
Price Index
         .767      -.212
.184
         Personal        1.04    Bank
         Income                  Reserves
    .433           .178
Total
Unemploymen
```

The correlation of the variables PINC and RESERVES is r=.95 (N=41, p=.00). Path diagram 2 confirms the importance of this relationship in defining the level of bank reserves.

Chapter Summary

The aging (movement) of foreign-born white immigrants throughout the U.S. economy appears to have an impact on the economy. Unlike the analysis of the current economic crisis, an approximate distribution of income by age is not available for analysis of its impact on the Great Depression economy.

The *beta coefficients* are the regression coefficients obtained by first standardizing all of the variables to a mean of zero (0) and standardized coefficient of one (1). These coefficients allow the relative contribution of each independent (predictor) variable to be assessed. Transforming the independent variables to standardized form makes the coefficients more comparable as they are in the same measurement units. For example, in Equation 1, z of the dependent variable (PINC) increases .071 for every change of one standard deviation IMMIGL16. In Equation 2, for every standard deviation of 1 z in PINC, RESERVES changes 1.04 z.

Table 12.5 Personal Income in Current Dollars (in billions), 1920-1940

Year		Year		Year	
1920	$73.4	1927	$79.6	1934	$54.0
1921	62.1	1928	79.8	1935	60.4
1922	62.0	1929	85.9	1936	68.6
1923	71.5	1930	77.0	1937	74.1
1924	73.2	1931	65.9	1938	68.3
1925	75.0	1932	50.2	1939	72.8
1926	79.5	1933	47.0	1940	78.3

Historical Statistics of the United States, Part I, p. 224, col. 8

In the end, what is occurring is a lack of effective demand. There is a housing boom (1922-1929) followed by a housing bust. The lack of effective demand is accompanied by an immigrant bubble and decline, unemployment, declining bank reserves, and a declining wholesale price index. This bubble and decline are accompanied by an increase, a decline, followed by a gradual increase and decline in personal income.

The Beta weights and further calculations make for the presentation of path analyses for both Equation 1 and Equation 2.

Path Analysis 1:

(1)	(2)	(3)
Variables in Equation 1 Path Analysis 1	Variation of 1 z in variable in column (1) yields variation of _____ in the standardized variable in column (3)	Variable (3)
Independent Variable		Dependent Variable
IMMIGL16	.071	PINC

WPI	.25	PINC
M_2	.807	PINC
WPI	.65	IMMIGL16

Correlation of WPI and M_2, r = .49, p = .002

Path Analysis 2:

TOTUNEMP	.178	RESERVES
WPI	-.212	RESERVES
PINC	1.04	RESERVES
WPI	.767	PINC
TOTUNEMP	.433	PINC

Correlation of TOTUNEMP and WPI, r= .18, NS

The path coefficients for IMMIGL16 (Equation 1) and PINC (Equation 2) both have positive, correct signs. The rapid increase in immigrants from 1900 to 1914, the rapid decline from 1914-1919, and the low levels of immigration lagged 16 years provide the lag in relation to the housing boom of the 1920s, the economic decline in 1929-1933, and the subsequent increase, decline, and increase in personal income. In Equation 2, WPI, and TOTUNEMP relate to PINC. Equation 2 leads to a superior equation and path analysis with variation in PINC accounting for almost all the variation in the dependent variable RESERVES.

Some explanation of the variable RESERVES is required. The United States had a system of fractional reserve banking during the Great Depression and still does. In such a system, money is created by two mechanisms. First, the banking system creates bank money, also known as "check book money". This is money created through the banking system by deposits, borrowing, and lending. The second type of money creation is central bank money. This encompasses all types of money created by the central bank (banknotes, coins, and electronic loans to private banks). If the money for a loan is supplied from central bank money new commercial bank money is created. When the commercial bank loan money is paid back the money disappears from existence as it is paid back to the central bank.

Check book money is the result of bank account holders depositing funds into a given bank. The bank sets aside the required amount for reserves, and then loans out the remainder. The remainder is loaned out to customers (individuals and businesses). They, in turn, deposit money into a bank. That bank sets aside reserves and loans the remainder of the money, and so on. To measure this increase in money supply a money multiplier is used. It calculates the maximum amount of money by which an initial deposit can be expanded with a given reserve ratio (the formula for the money multiplier is m = 1/R, R=reserves).

The decline of the reserves in the 1930-1933 period is thus dependent on two factors, central bank creation of money and checkbook money resulting from bank deposits from depositors. The amount of checkbook money became severely contracted by depositor withdrawals and bank panics in the 1930-1933 period (Wicker 1996). The Federal Reserve failed to increase money supply sufficiently thus reinforcing the reserve situation of banks (see Figure 12.8). The lack of liquidity of banks, increasing total unemployment, declining aggregate personal income, and hoarding all reinforced the decline in bank reserves.

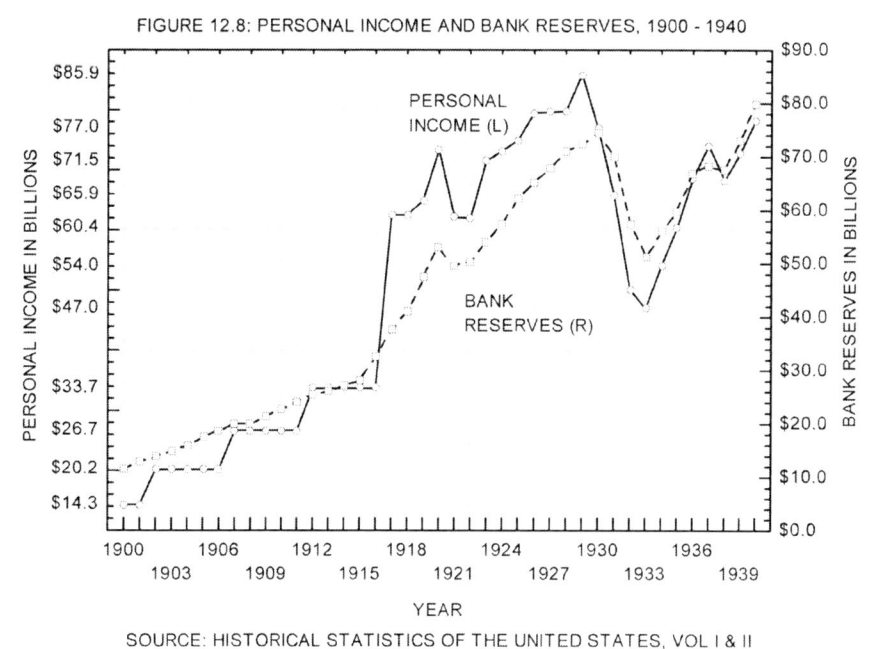

FIGURE 12.8: PERSONAL INCOME AND BANK RESERVES, 1900 - 1940

SOURCE: HISTORICAL STATISTICS OF THE UNITED STATES, VOL I & II

Part I, p. 224, col. 8, Part II, p. 1019, col. 581

In conclusion, the explanation of the variables leading to the Great Depression is multivariate in nature. PINC is the dominant variable in a path analysis (Path Analysis 2). The presence of a regression equation and a path analysis for RESERVES provides an empirical explanation of the essential variable RESERVES. Total unemployment and the wholesale price index partly explain the variation in personal income which in turn explains the variation in RESERVES.

Of importance to this argument is that calculating M_2 involves the use of checkbook money. Checkbook money has its origins in the deposits of depositors. It only partially reflects the actions of the Federal Reserve. M_2 reflects checkbook money during the Depression to a greater extent than it does central bank money. Thus, the actions of depositors in the 1930-1933 period were those of individuals having declining effective demand as represented by declining personal income. The decline in RESERVES was generated by declining numbers of individuals with declining deposits and little faith in the U.S. banking system.

Appendix I

Glossary of Abbreviations

GLOSSARY OF ABBREVIATIONS

AAA—Agricultural Adjustment Act; Agricultural Adjustment Administration

AFL—American Federation of Labor or A. F. of L.

CCC—Civilian Conservation Corps

CIO—Congress of Industrial Organizations

CWA—Civilian Works Administration

EBA—Emergency Banking Act

EHD—Emergency Housing Division

ERCA—Emergency Relief Construction Act

FCA—Farm Credit Administration

FDIC—Federal Deposit Insurance Corporation

FERA—Federal Emergency Relief Administration

FFB—Federal Farm Board

FHA—Federal Housing Authority

FLSA—Fair Labor Standards Act

FRA—Farm Relief Act

FSP—Food Stamp Plan

FSA—Farm Security Administration

FSRC—Federal Surplus Relief Corporation

GNP—Gross National Product

HOLC—Homeowners Loan Corporation

M_2—M_2 money supply

NCC—National Credit Corporation

NAM—National Association of Manufacturers

NIPA—National Income and Product Accounts

NLRA—National Labor Relations Act

NLB—National Labor Board

NLRB—National Labor Relations Board

NRA—National Recovery Act; National Recovery Administration

NRPB—National Resources Planning Board

NYA—National Youth Administration

NYSE—New York Stock Exchange

PINC—Personal Income

PUA—Public Utility Act

PWA—Public Works Administration

PWU—Project Workers Union

REA—Rural Electrification Administration

SSA—Social Security Act; Social Security Administration

TVA—Tennessee Valley Authority

UAW—United Auto Workers

UMW—United Mine Workers

URW—United Rubber Workers

USHA—United States Housing Authority

WPA—Works Progress Administration

WPI—Wholesale Price Index

WTA—Wealth Tax Act

WWI—World War I

Appendix II

An Equation for Aggregate Demand

Aggregate Demand	Personal Consumption	Government Capital Expenditures	Gross Domestic Private Investment	Net Exports
AD_{xt} =	PC_{xt} +	GOV_{xt} +	$GDPI_{xt}$ +	Z_{xt}

Testing Aggregate Demand

The relationships of the terms in the aggregate demand equation are presented here and tested.

Correlation Matrix (N=77)

	AD	PC	GOV	GDPI
PC	.999*	—		
GOV	.997*	.996*	—	
GDPI	.976*	.966*	.964*	—
Z	-.859*	-.873*	-.857*	-.837*

This relationship is a mathematical equality with the four independent variables and their coefficients adding to the term for aggregate demand. Thus, the adjusted R^2 = 1.00.

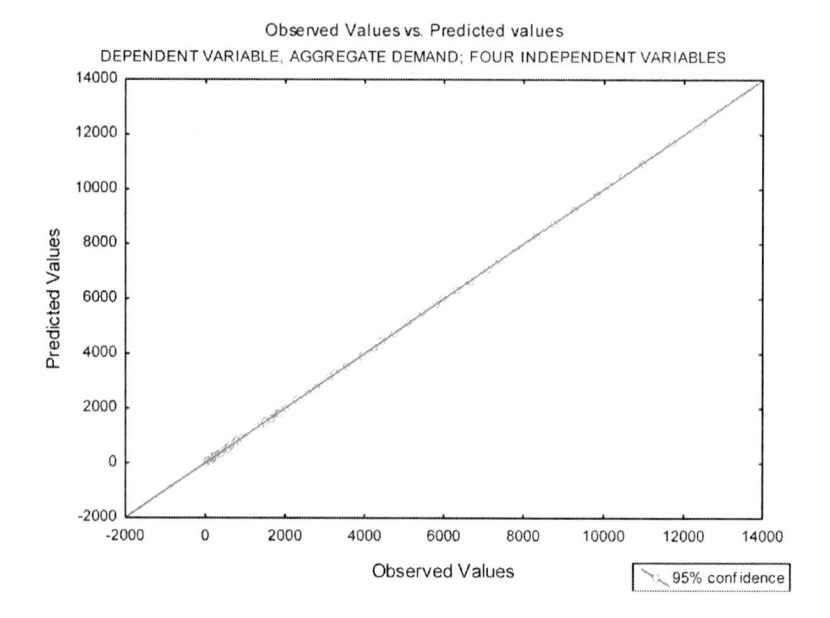

Observed Values vs. Predicted values
DEPENDENT VARIABLE, AGGREGATE DEMAND; FOUR INDEPENDENT VARIABLES

The following are the values for **Beta** in a standardized equation (N=77):

Variables	Beta
PC	.692
GOPV	.186
GDPI	.163
Z	.041

Means and Standard Deviations (N=77)

Variables	Standard Deviation	Mean
AD	3442.7	2766.3
PC	2382.1	1824.8
GOV	641.9	530.6
GDPI	559.5	472.7
Z	-61.9	142.0

REFERENCES

Age of Homeowner and Proportion Owning at that Age, *www.realclearmarkets.com*, accessed March 25, 2009.

Bell, Felicite C., and Michael L. Miller: 2009, "Life Tables for the United States Security Area 1900-2100, Actuarial Study 120", *www.socialsecurity.gov/OACT/NOTES/as120/LifeTables_Body.html*, accessed March 29, 2009.

Bernanke, Ben S.: 2000, Essays on the Great Depression, (Princeton University Press).

Bernanke, Ben S.: 1995, "The Macroeconomics of the Great Depression: A Comparative Approach", Journal of Money, Credit, and the Depression, XXVII, pp. 1-28.

Creamer, Daniel. Personal Income and Business Cycles. National Bureau of Economic Research, Princeton, NJ: Princeton University Press, 1956.

Current Population Survey, Annual Social and Economic Supplement (ASEC), Females: 2007, *http://pubdb3.census.gov/nacro/032007/perinc/new01_019.htm*, accessed March 23, 2008.

Current Population Survey, Annual Social and Economic Supplement (ASEC), Males: 2007, *http://pubdb3.census.gov/nacro/032007/perinc/new01_010.htm*, accessed March 23, 2008.

Department of Commerce, Bureau of Economic Analysis. National Income and Product Accounts, *http://www.bea.gov/bea/dn/nipaweb/*, list of all NIPA Tables.

Eichengreen, Barry and Jeffery Sachs: 1986, "Competitive Devaluation and the Great Depression: A Theoretical Reassessment." *Economics Letters*, XXII, pp. 67-71

Eichengreen, Barry and Jeffery Sachs: 1985, "Exchange Rates and Economic Recovery in the 1930s," *Journal of Economic History*, XLV, pp. 925-946.

Feldstein, Martin. *The Risk of Economic Crisis.* ed. by M. Feldstein. National Bureau of Economic Research. University of Chicago, Chicago. p. ix.

Fisher Irving. *Nature of Capital Income.* New York: Macmillan, 1906.

Fisher, Irving, and Harry G. Brown. *The Purchasing Power of Money, Its Determination and Relation to Credit, Interest, and Crises,* rev. ed. New York: Macmillan, 1926.

Fisher Irving. "Reflation and Stabilization," *Annals of the American Academy of Political and Social Science,* 171, no. 1 (1931): 127-137.

Friedman, Milton. "The Quantity Theory of Money—A Restatement," in *Studies in the Quantity Theory of Money*, ed. Milton Friedman, 3-21. Chicago: University of Chicago Press, 1956.

Friedman, Milton, and Anna Schwartz. *A Monetary History of the United States, 1867-1960.* National Bureau of Economic Research, Princeton: Princeton University Press, 1963.

Gibson, Campbell, and Emily Lennon. *Historical Census Statistics on the Foreign-born Population of the United States: 1850 to1990.* Working Paper No. 29, (U.S. Census Bureau, Population Division, Washington, D.C.)

Hall, Thomas E. and J David Ferguson. *The Great Depression, An International Disaster of Perverse Economic Policies.* The University of Michigan Press, Ann Arbor, 1998.

Hosen, Fredrick E. *The Great Depression and the New Deal, Legislative Acts in Their Entirety (1932-1933) and Statistical Economic Data*, Jefferson, NC: McFarland & Company, 1992.

Keynes, John Maynard. (The Galton Lecture delivered before the Eugenics Society, February 16, 1937.) "Some Economic Consequences of a Declining Population," *Eugenics Review*, 29, no. 1 (1937): 13-17.

Keynes, John Maynard. *The General Theory of Employment, Interest, and Money*. New York: Harvest/ Harcourt Brace & World, 1965.

Kindelberger, Charles P. *Manias, Panics, and Crashes: A History of Financial Crises*. rev. ed. New York: Basic Books, 1989.

Kondratieff, N.D.: "The Long Waves of Economic Life," trans. by W. F. Stolper, *The Review of Economic Statistics*, 17, no. 6 (November 1935): pp. 105-115.

Lebergott, Stanley. *Manpower in Economic Growth: The American Record Since 1800*, New York: McGraw-Hill Book Co., 1964.

Linder, Forrest E., and Robert D. Grove. *Vital Statistics Rates in the United States, 1900-1940*. Washington, D.C.: Federal Security Agency, United States Public Health Service, National Office of Vital Statistics, 1947.

Minksy, Hyman P. "The Financial Instability Hypothesis: A Clarification," in *The Risk of Economic Crises*, ed. Martin Feldstein, 158-166. Chicago: University of Chicago Press, 1991.

Mitchell, Wesley C., *What Happens During Business Cycles*. New York: National Bureau of Economic Research, 1951.

Mitchell, Wesley C. *Business Cycles: The Problem and Its Setting*. New York: The National Bureau of Economic Research, 1959.

National Association of Realtors. 2007, *http://homesandpeople.blogspot.com*, accessed September 17, 2008.

National Bureau of Economic Research. *http://www.nber.org/cycles.html*, February 2006.

National Center for Health Statistics. *http://cdc.gov/nchs/data/statab//4941x.01.pdf* Live Births, Birth Rates, and Fertility Rates by Race: United States, 1909-1994, accessed August 14,2007.

National Center for Health Statistics. Live Births, Birth Rates, and Fertility Rates by Race: United States, 1980-2004, *http://cdc.gov/nchs/data/nvsr/nvsr55/nvsr55_.01.pdf*, accessed August 14,2007.

Schupeter, Joseph A. *Business Cycles: A Theoretical, Historical, and Statistical Analysis of the Capitalist Process*, New York: McGraw-Hill, 1939.

Smith, B. Mark. *A History of the Global Stock Market, From Ancient Rome to Silicon Valley*, Chicago: University of Chicago Press, 2003

Social Security Life Tables for the United States Social Security Area, by Felicitie C. Bell and Michael L Miller. (Social Security Administration, Washington, D.C.). *http://www.ssa.gov/OACT/NOTES/pdf_studies/study120.pdf*, accessed March 23, 2008.

Temin, Peter. *Did Monetary Forces Create the Great Depression?* New York: W. W. Norton, 1976.

Temin, Peter. *Lessons from the Great Depression*, Cambridge, MA: MIT Press, 1989.

U.S. Department of Commerce, Bureau of Labor Statistics. Discouraged Workers, Seasonally Unadjusted, *http://www.bls.gov/legacy/cpsatab13.htm*, (accessed March 12, 2007).

U.S. Department of Commerce, Bureau of the Census. *Historical Statistics of the United States Colonial Times to 1970*, Parts 1 and 2. Washington, D.C.: Government Printing Office, 1975.

U.S. Department of Commerce, U.S. Census Bureau, Population Division, Campbell Gibson and Emily Lennon. *Historical Census Statistics on the Foreign-Born Population of the United States: 1850-1990*, Working Paper No. 29, 1999.

United States Department of Commerce, Bureau of Economic Analysis. National Income and Product Accounts, *http://www.bea.gov/bea/dn/nipaweb*, select all tables, (accessed March 1, 2007).

United States Statutes at Large. Part I, April 1921-March 1923, Chapter 8, 67th Cong., 1st session, p. 5, and 2nd session, p. 540. Washington D.C.: Government Printing Office.

Wecter, Dixon. *The Age of the Great Depression: 1929-1940*, New York: Macmillan, 1948.

Wicker, Elmus. *The Banking Panics of the Great Depression,* Cambridge University Press,. Cambridge, 1996.

INDEX

A

AAA payments, 77
accumulation, 105
adjusted unemployment, 173, 177-78, 207, 211, 229
Agricultural Adjustment Act (AAA), 59, 75, 77-78, 239
Agricultural Adjustment Administration, 155, 239
American Federation of Labor, 64, 239

B

Bankhead-Jones Act, 73
banking
  bank closures, 39
  bank failures, 23, 40, 45
  bank reserves, 40
  banks, declines in numbers, 182
  national banks, 53-54
  non-national banks, 8, 174-77
  small banks, 45
Bernanke, Ben, 35-36, 160-61

C

capitalism, 96-97, 107, 125, 145-46
Civilian Conservation Corps (CCC), 57, 83, 239
civilian labor force components
  forty-five-year-olds, 25-27, 29
  sixty-four-year-olds, 25-27, 29
Civilian Works Administration (CWA), 58, 239
company-sponsored unions, 64
Congress of Industrial Organizations (CIO), 66, 81, 239
consumer buying power
  decline in, 44, 183
consumer price index, 157-58, 183-84, 211
Crash, 19-20, 32, 37-38, 41, 45-47, 219, 224
creative destruction, 20

D

decumulation, 105
deflation, 30, 34-35, 159
  consumer price index, 157-58, 183-84, 211
deposits, 126, 170-71, 180-81, 202
dissaving, 105
Dust Bowl, 80-81

E

economic evolution, 33-34, 105, 107, 110, 113, 115-16, 118-19, 121-22, 127, 145
effective demand, 162

253

Eichengreen and Sachs, 35-36, 159-61
Emergency Banking Act (EBA) of 1933, 39, 54, 130, 132, 153, 239
Emergency Housing Division (EHD), 68, 239
Emergency Relief and Construction Act (ERCA), 129, 239
employment, 61-62, 87, 162
 farm, 71, 142, 185-87
 nonfarm, 71, 142, 185-86
equilibrium, 33, 102-4, 117-21, 148, 162-67
 disequilibrium as opposed to, 101, 116-17
 unstable equilibrium and, 100
estimation
 determining loss in personal income through, 37
 determining number of discouraged workers through, 196
 determining of M2 through, 172
exports, 47-48, 143-44

F

Fair Labor Standards Act (FLSA), 67, 82, 239
Farm Credit Administration (FCA), 75, 239
farmer revolts, 75
farming, 6, 20, 67, 69, 76, 78, 80, 105, 134, 185, 187
 types and numbers of farms for, 72
farming methods, 76
farm mechanization, 96
Farm Mortgage Corporation Act of 1933, 132

farm mortgage moratorium, 64
farm ownership, 72, 189
Farm Relief Act (FRA) of 1933, 132-33, 239
Farm Security Administration (FSA), 69, 73
Federal Deposit Insurance Corporation (FDIC), 42, 135, 137, 202, 239
Federal Emergency Relief Administration (FERA), 57-58, 80-81, 84, 87, 239
Federal Farm Board (FFB), 74, 239
federal government, role changes in, 67
Federal Housing Authority (FHA), 68, 239
Federal Surplus Relief Corporation (FSRC), 76, 240
Food Stamp Plan (FSP), 76, 240
Friedman, Milton, 36-37, 46, 156, 170-71
Friedman-Schwartz money supply hypothesis, 36

G

Glass-Steagall Act, 55, 129
gold hypothesis
 Bernanke's view of, 35, 160
 Eichengreen's and Sachs's view of, 35-36, 159-60
 Temin's view of, 34-35, 149, 151, 157, 159
gold standard, 34-36, 151, 153, 156-57, 159-61
government capital expenditures, 208
Great Depression
 bank reserves during, 40

conceptual view of, 92, 94
crime during, 82
declining personal income during, 17, 23, 38, 42, 131, 170, 199, 228-29, 232-33, 235, 240
defense needs at the end of, 67
foreclosures during, 38, 40, 68, 225
housing boom and bust during, 38, 40, 51, 221
lack of faith in financial institutions during, 23
liquidity and illiquidity during, 170
M2 level during, 23
new perspectives on, 41
unemployment during, 23, 38, 40-41, 47, 61-62, 64, 79, 150-51
gross domestic private investment, 195, 207-8
gross national product (GNP), 178-80, 211, 240

H

Hawley-Smoot Tariff Act, 47
Home Owners' Loan Acts of 1933 and 1934, 132
Home Owners' Loan Corporation (HOLC), 68
Hoover, Herbert, 49-53, 57, 64, 74, 91
housing (nonfarm)
  foreclosures, 221-22, 224-25
  starts, 38, 219-24
hunger marches, 49

I

immigrants
  increase in 1900 to 1914 of, 37-38, 217, 219, 225-26
  low levels in 1915 to 1919 of, 217
immigration, 17, 37, 41, 185, 216-17, 219
imports, 47, 126
incomes, consumer, 85
innovation, 106-19, 121
interest, 112-14, 116

J

Jugular wave, 33, 122, 148

K

Keynes, John Maynard
  effective demand as defined by, 162-66, 168-69
Kitchin wave, 33, 122, 148

L

life expectancy, 225
live births
  decline in 1961 to 1975 of, 24
  increase in 1933 to 1960 of, 24
  survival rate multiplied by number of, 26
loans, 68, 113, 182-83

M

M2 money supply, 201, 228
market Crash, 19, 32, 37, 41
money supply, 36, 123, 144
  types of
    central bank money, 233
    checkbook money, 38, 234-35
    M2, 171-73, 177-80, 183, 199-203, 235

mortality, 81, 226, 229
multiple regressions with ß, 229, 231

N

National Association of
　Manufacturers (NAM), 59-60,
　65, 240
National Credit Corporation (NCC),
　50, 240
National Education Association, 49
National Income and Product
　Accounts (NIPA), 207, 240
national income decline, 47
National Labor Board (NLB), 65, 240
National Labor Relations Act (NLRA),
　60, 64, 240
National Labor Relations Board
　(NLRB), 65, 240
National Recovery Act (NRA), 59-60,
　65, 82, 133-35, 240
National Recovery Administration, 240
National Resources Planning Board
　(NRPB), 63, 240
National Youth Administration (NYA),
　84, 240
New York Stock Exchange (NYSE), 32,
　240
1938-1939 recession, 89

P

path analyses
　path diagram 1 and ß of, 228-30,
　232
　path diagram 2 and ß of, 231
Patman Bill, 49
personal consumption, 17, 30, 38,
　41, 208-11, 229

personal income (PINC), 17, 37-38,
　42, 158-59, 161, 170, 182, 193,
　199, 210-14, 226-29, 232-33,
　235, 240
　growth and decline of, 195, 210
population decline, 80
population subgroups, 161
profits, 111, 114
prohibition repeal, 57
　local-option laws during, 60
Project Workers Union (PWU), 66,
　240
Public Utility Act (PUA), 60, 240
Public Works Administration (PWA),
　57-58, 67-68, 91
public works and relief, 57

R

Railway Labor Act, 63
reserves, 144, 146, 157, 229
Resettlement administration,
　medical care during, 86
Revenue Act of 1936, 60
Robinson-Patman Act of 1937, 59
Roosevelt, Franklin Delano, 34, 39,
　48, 51-52, 147, 153-54
　first term of, 52-57, 59-60
　second term of, 62-64, 77, 80,
　87-91
Rural Electrification Administration
　(REA), 78, 241
rural to urban movement, 67

S

sales tax, 60
Schumpeter, Joseph, 20, 33, 97-101,
　104-8, 112-18, 122-23, 137-44

capitalist evolution as defined by, 33, 95-96, 109, 125-26, 145-47
the Depression according to, 125-29, 131-33
recovery according to, 34, 119-21, 130-35, 140, 240
Schwartz, Anna, 36, 46, 170-71
Social Security Act (SSA), 60, 64, 81, 87, 89, 241
South, 69, 72, 74, 77-79
strikes, 66-67
Supreme Court, 59, 63-64, 76, 134
survival rates, 217-18

T

Temin, Peter, 34-35, 149, 151, 157, 159
Tennessee Valley Authority (TVA), 58, 88-89, 241
theses, 193-94, 209

U

unemployment, 38, 40-42, 47, 61-62, 67, 79, 149
adjusted, 173, 177-78, 207, 211, 229
United Auto Workers (UAW), 66, 241
United Mine Workers (UMW), 64, 241
United Rubber Workers (URW), 66, 241
United States, precrash, 44-45

V

velocity, 178-80, 214

W

Wagner Labor Relations Act, 64-65
Wagner-Steagall Act
low-cost housing as part of, 68
United States Housing Authority (USHA) under, 68, 241
Wealth Tax Act (WTA), 60, 241
wholesale price index (WPI), 42, 137, 170, 199-201, 203, 211-14
Wicker, Elmus
bank panics according to, 38-40
worker's compensation laws, 60
Works Progress Administration (WPA), 60, 66, 78, 90-91, 241
accomplishments of, 61-62
World War I (WWI), 34, 49, 151
debts incurred for, 48

Y

youth unemployment, 82

HB 3722 .S255 2010

Salisbury, Philip S.

The Current Economic Crisis
and the Great Depression